DISASTER:
RECOVERY

**Your personal guide to
surviving the first few weeks
of a natural disaster**

Woodslane Press Pty Ltd
10 Apollo Street
Warriewood, NSW 2102
Email: info@woodslane.com.au
Tel: 02 8445 2300 Website: www.woodslane.com.au

First published in Australia in 2018 by Woodslane Press
© 2018 Woodslane Press, text © 2018 Collyn Rivers

A catalogue record for this
book is available from the
National Library of Australia

NATIONAL
LIBRARY
OF AUSTRALIA

Book design by: Jenny Cowan, Collyn Rivers and Max Nolan

CONTENTS

PART FIVE: EXTRAS

REFERENCES & DATA: (CLIMATE ETC)

REFERENCES & DATA: (MEDICAL)

REFERENCES & DATA: (GENERAL)

PREFACE

Disaster : Recovery is a down-to-earth book that explains how you and your family can cope during and particularly after disasters that *can and do* happen (not just *may* happen) around the world at any time.

The after-effects of such disasters are similar worldwide. Even though various countries use differing terms this book thus suits global readers. Here's just a few recent examples.

August 2005 saw Hurricane Katrina all but destroy New Orleans. In 2011, Queensland (Australia) suffered devastating floods, followed days later by Cyclone Yasi. It was the strongest cyclone ever recorded in Australia.

In 2010, Christchurch, New Zealand, suffered a strong earthquake that badly damaged many buildings. Less than two years later, an even more serious earthquake destroyed most of its remaining infrastructure.

The catastrophic earthquake and tsunami experienced in Japan in 2011 destroyed a massive area. It left a huge number of people homeless and without basic services for many months.

In recent years, Australia, Chile, Israel, New Zealand, Spain and the USA experienced major bush and forest fires: many worse than in living memory. Experts predict such fires will become more frequent and intense.

PREFACE

SIMILAR GLOBAL EFFECTS & NEEDS

Water and electricity usually fail, often for weeks or more. Telephones stop working. Roads are impassable or closed to all but emergency and essential traffic. Food shops and fuel stations run out of supplies.

Police, paramedics and other emergency services do their best but lack resources to address any but truly serious issues.

As emergency services must scale priorities, individual or family survival may depend substantially on personal effort – and knowing how.

Despite the major and known effects of the *aftermath* of disasters such as fire, tsunami or earthquake, little has previously been written about the following weeks.

This book mainly covers urban scenarios: houses, apartment buildings, tower blocks and low-lying areas. It assists country folk too (even though most are accustomed to coping when things go badly wrong).

As many people now travel extensively in recreational vehicles, the book also shows how to cope with their major risk: of severe wind events.

NOT JUST PHYSICAL

The benefits of preparation are not just physical. Even minor preparations greatly increase a sense of wellbeing post-disaster.

People who make even limited preparations cope better physically and psychologically than those who do not.

Those who do not (prepare) suffer, not just through a lack of basic vital needs; they feel helpless as a result.

MEDICAL INFORMATION

A major issue during disasters and post-disasters is that professional medical services tend to be overwhelmed or non-available.

Medical advice, however, can only be given legally by qualified medical professionals. It is, however, legal to provide medical *information.*

It is thus medical *information* only (*not advice*) that is included in this book. It is intended only for last resort use in emergency situations where professional medical advice and attention absolutely cannot be obtained.

Do not disregard or delay seeking professional medical advice because of something you have read in this book.

The medical content is sourced from public domain professional medical material (from the USA, Canada and Australia). No warranty is given that it is complete, current or correct. It is not medical *advice* of any kind.

FIRE ARMS

Fire arm usage and legislation varies from time to time and country to country. See page 118 for references.

PART ONE

The Risks

The first two weeks
Bush fires
Earthquakes
Floods
Wind events
Re-entering damaged buildings

THE FIRST TWO WEEKS

No matter the cause, the major problems encountered post-disaster are surprisingly similar.

The most common is avoiding disease from contaminated water, yet obtaining water sufficiently clean to avoid serious dehydration.

This problem almost always follows floods and earthquakes but also, and unexpectedly, after the 2011 nuclear reactor disaster in Japan.

VITAL ISSUES

There are many other vital issues. The most important include:

- *Contacting others without working phone lines or mobile phones.*
- *Cooking without electricity or mains gas.*
- *Coping without a working toilet.*
- *Eating safely and adequately.*
- *Finding out what's going on locally and nationally, and how and what others are doing.*
- *Keeping well – as medical care may not be available.*
- *Knowing when it's safe to re-enter damaged buildings.*
- *Shopping without normal transport.*
- *Staying warm or cool.*

No matter where in the world you live, or what catastrophic event you experience, you are almost certain to be faced with some or all of the above issues.

This book shows you how to cope.

BUSH FIRES

Known variously as wildfires, bush fires, brush fires etc, these occur in places as diverse as Australia, Brazil, France, Greece, Japan and Russia, and particularly in California (USA).

By early May 2011, wildfires had already burned nearly 2.4 million acres in the United States. It was the most active and intense early fire season in a decade.

Whilst the numbers killed on average each year are low, the data conceals the *local* intensity of risk.

One, in the Landes forest in France (in 1949) killed 83 people including firefighters.

Australia's bush fires typically kill one or two people each year but one (in Victoria – in 2009) killed 173.

FIGHT OR FLEE?

There are two different views on whether to stay: fight the fire and protect your home, or flee in ample time.

The *Los Angeles Times* summed this up in an article (August 2008): 'Fire is a pervasive danger in Australia .... but Australians cope in a manner difficult to envisage in the USA.'

'Americans expect fire fighters to protect their lives and property. Australians in rural communities view that as their own responsibility.'

The rationale of staying and fighting fires is that houses are mostly ignited by fiery embers *after* the fire front has passed.

Whilst fleeing may seem safer, many attempting to do so during the 2009 Victorian bush fires died in car crashes. Cars and their occupants were also trapped by fallen trees, power lines – or by the fire itself.

Australia's experience is that unless fleeing very early, the risks of belated fleeing are higher than of staying.

National standard fire danger rating scale alerts, and forecast danger levels, now exist in many countries. These help guide appropriate responses: see page 104.

PRE-PLANNING AND PREPARATION

This book summarises basic and vital fire issues – sourced primarily from American, Australian and Canadian fire authorities.

As fire behavior varies considerably from area to area check the references on page 104 for local information.

Also check out 'wildfire ready' community initiatives that encourage and support people to form prepared groups.

PREPARING YOUR HOME

Whether fleeing or staying, preparing the property greatly increases its likelihood of surviving all but catastrophic fires. It also eases the work of fire fighters.

In many major fires, some homes burn while others do not. Of those that burn, some are ignited as the fire front passes over them, but many by wind-blown embers that ignite anything flammable they land on.

For country properties, mark out a (minimum) 20 metre (22 yards) ring around your home and other buildings. Keep *everything* inside that ring clear of anything that readily burns, e.g. debris, dry grass or old tires. Remove all flammable fluids and materials and store them well away from the main house.

Create fire breaks by clearing a three yard (metre) strip along property boundaries (this is obligatory in some areas). Prune all branches below head height and ensure none are within two to three yards of buildings.

Remove anything flammable lying against buildings, such as firewood or garden mulch. Clear all twigs and litter. Relocate any gas cylinders to more fire secure locations. Install metal fly mesh across all doors and windows.

Block access under eaves, roofs, skylights, external air conditioners, floors or anywhere that may trap or admit flying sparks or embers.

This is particularly important on walls likely to face an approaching fire front.

Have a readily accessible supply of water — such as a tank, dam or swimming pool — plus a diesel or petrol-powered water pump of adequate size.

Contact your local fire brigade to check that its hose connections match yours on the water tank (if you have one). Use large-diameter canvas hoses – again ensuring connections match. Ensure all taps and connectors are metal so they won't melt during a fire.

Commercial gel fire retardants are claimed to assist. A mixture of four litres (approximately one gallon) with 200 litres (approximately 44 gallons) of water covers about 100 square metres (120 square yards) of building.

BUSHFIRE BUNKERS

Various homemade and (usually) underground commercial fire bunkers are available. These are mainly from the USA, Canada and Australia.

There is official concern that not all are safe.

Victoria's (Australia) Bush Fires Royal Commission found that while existing bush fire shelters may well save some people, they may kill others.

Australia has an official guide that greatly assists. It can be downloaded (as a free .pdf) from abcb.gov.au/Resources/Publications/Education-Training/Private-Bushfire-Shelters%20

The guide recommends to have only an above-ground or an in-ground shelter *separated from* an associated dwelling. It emphasises that fire risks to life in shelters such as a cellar or a 'safe' room within a dwelling are considerable – and too great to accept.

The guide stresses that a shelter that is built as described provides 'a measured degree of protection to people with nowhere else to go, such as occupants of dwellings in remote areas – but that such a shelter must be a package of measures that, combined, form a robust 'Bushfire Risk Management Strategy'.

LIVESTOCK AND PETS

Ideally, slash paddocks prior to the fire season, otherwise move animals to your clearest paddock.

Do not allow livestock or pets onto public roads. In their panic they may hinder emergency services.

Few public shelter and relocation centres accept pets. Try to prepare a safe area and leave ample food and water.

WHAT TO EXPECT

Live fallen power lines are likely to be encountered if fleeing through just previously-burned areas.

Electricity seeks a path to earth – and that path may be a tree, a vehicle, fence or anyone seeking to assist.

Keep away from *any* object in contact with a power line. The voltage is very high at the point of electrical contact. Farther away, the voltage drops off.

With power lines of up to 60,000 volts, the voltage typically drops to zero about 10 metres (11 yards) away.

Why fallen power lines can be so dangerous is that if you were to stand side-on (i.e. such that one foot is closer than the other to the point of electrical contact) you may have several thousand volts difference between your feet.

Electric current will pass up one leg and down the other via your heart – killing you instantly.

Always assume that electric cables are live.

Unless there are sparks or movement you cannot tell if a power cable (or the ground it lies on) is 'live'.

Even if initally 'dead', power may suddenly be automatically restored. There may well be no physical indication that this has occurred.

EARTHQUAKES

The earth's surface can be perceived as a few large plates, and many small plates, averaging each about 80 km (50 miles) thick.

They move from zero to about 100 mm (4 inches) a year. This causes stress to build up and cracks (fault lines) to form within the plates.

Every now and again a plate moves suddenly. This results in sudden shocks that affect the earth's surface. It is nature's way of releasing stress.

Earthquakes may occur almost anywhere but generally above fault lines. There is still no reliable way to predict when an earthquake will occur in any given location.

In the more earthquake-prone areas, however, it is not a matter of *'if'* an earthquake might occur. It is a matter of *when* it will occur.

EARTHQUAKES: THE RICHTER SCALE

The Richter scale indicates the 'size' of the related earthquake shock wave.

Each Richter unit represents a 10 times higher shock wave but over 30 times more of the energy resultant damage.

A Richter 9 earthquake is thus over 900 times more severe in its effects than one of Richter 7.

An earthquake exceeding 5.5 or so on the Richter scale is likely to wreck electric, gas, water, sewerage, and telephone lines.

It may also affect television, radio and mobile telephone towers, weaken or destroy bridges, may damage or wash away roads and cause airport closures.

Earthquakes may kill people during the main event, and during ongoing aftershocks. They cause massive long term damage to services and infrastructure.

Extensive US, Japanese and New Zealand experience, however, indicates that existing structures can be made safer and less prone to damage.

PRE-PLANNING

There is much you can do to prepare for earthquake impacts. This greatly increase your chances of survival if you are at home when an earthquake occurs.

At the very least, pre-planning and preparation may save your house. This will help you and your family during the aftermath.

PREPARING YOUR HOUSE

The strongest earthquake protected buildings are of well constructed, diagonally braced timber, securely tied to adequate foundations and topped with light timber or metal roofs.

Steel framed buildings too need diagonal bracing.

Brick veneer is more flexible and earthquake resistant

than double brick and stone. Should you live in a brick, stone or brick veneer house, check for unsupported masonry parapets, gables and chimneys. Repair loose roof tiles and any cracks in the walls.

Dwellings most at risk are of un-reinforced concrete, masonry or brick. Chimneys are particularly vulnerable but it is often possible to add support. Also see 'Entering damaged buildings' on pages 33-38.

FURNITURE AND FITTINGS

Heavy furniture (such as bookcases) located against walls should be attached firmly to those walls. Have large ornaments or heavy objects only on low shelves.

Cupboards containing heavy objects should have safety latches. Items such as water heaters must be securely fastened.

Lock any castors on heavy items, such as barbecues.

INSURANCE

Check that your insurance policy covers earthquake damage (not all do) and that you are fully insured.

DURING AN EARTHQUAKE

When an earthquake strikes, unless you are outside it is generally safer to stay wherever you happen to be at the time.

If outside, move well away from buildings of all kinds. Most deaths and injuries occur to people entering or leaving buildings as masonry crashes down.

If inside a building, *stay there* and apply the following procedures.

• *Keep well away from windows, mirrors, and glass in general.*

• *Check there is nothing overhead such as fluorescent lights that may crash down.*

• *Stay clear of heavy furniture such as bookcases that may topple on to you.*

• *Stay clear of any form of heating unit.*

• *If possible, take cover under a sturdy table or bed – or whatever you can find to protect your head (in particular) from falling debris.*

• *If you can't find cover, brace yourself in an inside corner of load bearing walls but not partition walls – and definitely not a chimney as masonry may fall from it.*

• *Open all exit doors if possible and them leave open as they commonly jam closed as a building distorts.*

• *Do not use elevators.*

• *Be especially wary of staircases. They can inflict terrible injuries if you are caught on one as it collapses.*

Once you have located a safe place, don't leave it. Move with caution while carefully assessing the situation.

DROP, COVER AND HOLD

The Drop, Cover and Hold technique (shown below) is recommended by official earthquake authorities worldwide. It applies particularly for those caught in a house or office building.

DUCK or DROP on to the floor.

Take **COVER** under a strong desk or table. If impossible, move against an inside wall. Use your arms to protect your head and neck. Avoid windows, mirrors, hanging objects, bookcases and other tall furniture.

If under a strong shelter, **HOLD** on to it tightly and if necessary move with it. Hold until shaking ceases and it is safe to leave.

'TRIANGLE OF LIFE' – CAUTION

The 'Triangle of Life' concept advises that earthquake wreckage often has triangular voids that are safer to be in than anywhere else.

The official United States Geological Survey agency, however lists this as an 'urban myth'.

It states: 'The Triangle of Life' is a misguided idea about the best location a person should occupy during an earthquake'.

According to the US Red Cross: '**DROP, COVER, and HOLD**' under a table or desk is still the best recommendation'.

AFTER THE EARTHQUAKE

Initial shocks rarely last more than a minute or so. They are almost always followed by a series of shocks that are usually (but not invariably) smaller.

Don't immediately get up and run when the initial earthquake seems over. Aftershocks are virtually certain.

RISK OF FIRE

Fire is a common consequence of earthquakes. If you can, turn off the gas at the main tap.

Do not light any matches until you are 100% certain that no gas is leaking.

FLOODS

Floods occur worldwide with those recorded in Australia since 1852 claiming at least 950 victims, and one in southeast Queensland causing 35 confirmed deaths as recently as 2010-11.

The main causes of flooding are:

Seawater – particularly after a cyclone or severe storm that coincides with a high tide and/or higher river levels; also tsunamis following undersea earthquake.

Run-offs – may follow a dam overflow or excess rain or melting snow.

Faulty or overladen drainage – this is an urban issue than can be unexpected, swift and severe.

Unlike most natural disasters, most die whilst attempting to flee – typically underestimating the forces involved when flood water flows in volume.

PRE-PLANNING

Local councils have flood maps and plans. These show problem areas, escape routes and the location your closest relief centre.

There is usually ample warning with regular radio updates. This gives plenty of time for packing warm clothing, essential medications, papers and mobile phone etc.

Before leaving place sandbags in all toilets and drains
– unless done sewage is likely to back-flow into your
home. Let a relative or friend know where you are
going – and how. Ask that if they do not hear from you
by an agreed time, to contact emergency services.

If possible avoid driving or walking in flood water,
but if you have to wear strong shoes. Use a walking
stick or a broomstick and constantly check the water's
depth and flow. Do not enter any flood water more than
knee-high. If caught out in any way call for assistance.

Do not return home until you are officially advised
that it is safe to do so.

POST FLOOD

As with so many disaster situations, the immediate
time after presents many unexpected dangers.

The aftermaths of a major flood are generally similar to
those of other natural disasters but may last longer and
affect a much wider area.

There is likely to be widespread loss of electricity
and a lack of food and water. Roads may be closed
to all traffic except emergency vehicles. Medical and
essential services are not likely to be available for
some days.

Groundwater is almost certain to be contaminated
and may continue to be for many weeks afterwards.

FLOODS

A tsunami is a massive surge often caused when an earthquake's epicenter is out to sea. It lifts a huge expanse and mass of water by a metre or so.

There have been over 50 recorded tsunamis affecting Australia's coastline since European settlement. In 1977 a six metre tsunami travelled inland at Cape Leveque, WA. In 1994 a tsunami travelled 300 metres inland in the Onslow-Exmouth region of WA. In 2006 a tsunami affected parts of the WA coast, particularly at Steep Point where a tsunami travelled 200 metres inland.

Japan's 2011 tsunami had a peak of 14 metres (15.3 yards).

The initially shallow wave moves at colossal speed: it can cross the Pacific in less than a day.

Tsunamis typically slow to 40 kilometres/hour (25 miles/hour) as they encounter less deep water.

The waves rarely break but their height increases. They penetrate a long way inland in a deep steady flow that destroys all in its path as they advance.

The surge may extend up to 10 kilometres (about six miles) inland and cause far more damage than the originating earthquake. It then sweeps vehicles, houses, loose debris and people out to sea as it retreats.

How to cope after floods and tsumanis is covered on pages 40 onward.

WIND EVENTS

Cyclones, hurricanes and typhoons are different names for generally similar wind events - excepting that 'typhoon' relates only to strong cyclones in the western North Pacific basin. (To avoid duplication, most such are referred to in this book as 'cyclones' or wind events.)

Most cyclones travel at least as fast as humans can run. They usually gain strength as they do so, but the *slower* they travel the more powerful they become.

All such events wreak a path of devastation, killing and seriously injuring people and destroying or badly damaging buildings, boats and vehicles.

Excepting that no cyclone is fully predictable, they may travel great distances before typically crossing a coastline. Some may back-track for a day or two – or change direction.

All eventually lose strength and die, particularly if they cross cooler water.

KNOWING THERE'S A CYCLONE COMING
There's rarely visual signs of an oncoming cyclone excepting that coastal dwellers may see frigate birds flying away from it. Snakes and other reptiles appear to sense cyclones and may be seen seeking shelter.

More reliably, cyclone-affected countries have their own warning systems.

WIND EVENTS

The National Hurricane Centre's Hurricane Specialist Unit constantly watches for tropical cyclones and indications within the North Atlantic and eastern North Pacific basins.

The Centre provides information for the USA and the Caribbean, plus world meteorological organisations. See pages 109-110.

CYCLONE WARNINGS (AUSTRALIA ETC)

Australia's Bureau of Meteorology (BoM) warns of approaching cyclones on regular news broadcasts on local radio, via State Emergency Services, an automated telephone service and via bom.gov.au/cyclone/.

The BoM issues 'Tropical Cyclone Advice' whenever a tropical cyclone is expected to cause winds in excess of 62 km/h (gale force) over land in Australia. The advice is either a 'Watch' (the onset of gales is expected within 48 hours but not within 24 hours) and/or a 'Warning' (issued for coastal communities when the onset of gales is expected within 24 hours or is already occurring). BOM have a Current Tropical Cyclones page with detailed information on the current position, intensity and expected track of any cyclone in Australia.

Since cyclone response systems, evacuation procedures and cyclone shelter locations vary considerably according to local conditions, educate yourself about these for your area well before a cyclone hits.

Queensland, Northern Territory and Western Australia provide detailed information on this. If you are in a caravan park in an affected area, the park will have its own evacuation procedures and trained staff. Study these in good time and follow the instructions of staff.

Finally, it is also important to use common sense as well as your own eyes and ears in such situations. As noted by the Federal Government's Solicitor General in 2013:

It is important that communities do not rely solely on receiving an alert or emergency warning as warnings may not be able to be issued in all circumstances. People should not ignore their own observations and visual cues that alert of an approaching hazard. Individuals should be prepared and have an action plan in case of an emergency.

WHEN A CYCLONE IS CERTAIN

• *Have a battery radio tuned to the local news broadcast.*

• *Check the web using the addresses above and/or in the References: pages 108-110.*

• *Ensure your immediate emergency gear is close by.*

• *Unplug all electrical appliances.*

• *Stay in the strongest part of the building.*

• *Identify a safe place to shelter during the cyclone. The toilet, or beneath a building's stairs, tend to be strong areas. If you cannot make at least one place strong enough, it is best to find safe shelter elsewhere.*

WIND EVENTS

WHEN A CYCLONE STRIKES

• Do not attempt to look outside.

• If the building begins to break up, protect yourself with mattresses, ideally under a strong table or on the floor.

• If things go quiet, remember that cyclones have a neutral hole in the center. That quiet is likely to be the hole passing overhead. If so the cyclone will build up again shortly.

TYPICAL GRADED WARNINGS

A Cyclone Watch (or similar) warns if cyclone activity develops within a typical 100 kilometres (62 miles) of any area that might be affected.

If a wind event continues to grow, it formally becomes a cyclone, hurricane etc and is given a name (e.g. 'Amy'). Graded warnings (much as below) are then provided.

BLUE ALERT: A cyclone may affect the area but strong winds are not yet a threat. Move to a safe shelter or prepare the place where you intend to stay if you have not done so already.

YELLOW ALERT: A cyclone is moving close and appears certain to continue. Strong winds are predicted. Police and safety workers are likely to close off all roads (except to emergency vehicles).

If you do not have adequate shelter, seek help as soon as possible.

WIND EVENTS

RED ALERT: A cyclone and dangerous winds that cause damage are about to hit your area. Seek urgent help now unless prevented by wind forces or flooding, in which case look for the strongest shelter you can find that is well above sea level.

ALL CLEAR: This signals the end of the **Red Alert.** It is safe to go outside but proceed with caution because a **Yellow** or **Blue Alert** may still be current.

WIND SPEEDS AND PROBABLE DAMAGE

There is no universal scale for cyclone strength nor do all scales indicate similar effects.

The category of a typhoon is decided by the maximum sustained winds, but a typhoon in the Japanese standard and a typhoon in the international standard are not the same: pages 111-113.

Each doubling of wind speed increases forces by four times. A 300 km/h (180 miles/hour) wind has nine times the force of a 100 km/h (about 60 miles/hour) wind.

THE AUSTRALIAN/FIJIAN CYCLONE SCALE

This scale ranks cyclones based on wind gusts averaged over three seconds - and probable damage caused.

Category one: (tropical cyclone) < 125 km/h (78 miles/hour). Negligible house damage. Damage to crops, trees and caravans. These winds correspond to Beaufort 8 and 9 (Gales and strong gales).

WIND EVENTS

Category two: (tropical cyclone) 125-164 km/h (102 miles/hour). Minor house damage, significant damage to signs, risk of power failure. Heavy damage to some crops. Small craft may break moorings. These winds correspond to Beaufort 10 and 11 (Storm and violent storm).

Category three: (severe tropical cyclone) 165-224 km/h (111-130 miles/hour). Roof and structural damage. Caravans destroyed. Power failure likely. A Category 3 cyclone's strongest winds are very destructive. These winds correspond to the highest category on the Beaufort scale, Beaufort 12 (Hurricane). An example is Tropical Cyclone Roma.

Category four: 225-279 km/h (140-173 miles/hour). Significant roof loss & structural damage. Many caravans destroyed and/or blown away. Power failure highly likely. An example is Tropical Cyclone Tracy. These winds correspond to the highest category on the Beaufort scale, Beaufort 12 (Hurricane).

Category five: greater than 280 km/h (174 miles/hour). Extremely dangerous with widespread destruction. Power outages may last for weeks. A Category 5 cyclone's strongest winds are very destructive with typical gusts over open flat land of more than 280 km/h.

(Australia's and Fiji's scale are based on peak wind gusts.) Readers in other cyclone-prone areas should check their regional or national centres (pages 108-109).

WIND EVENTS

TORNADOS

Tornados are very rare in Australia but have occurred. They develop inside severe rotating thunderstorms ('supercells') as inverted funnel-like clouds that extend finally to the ground.

Tornados are about 15 km (about 10 miles) across at the top and taper as they descend. They usually move at around 50 km/h (about 30 miles/hour). Most last only minutes: it is a rare tornado that lasts for hours. As with cyclones, tornados are massively destructive. Their concentrated power destroys whole communities as they cross.

AFTER THE EVENT

The post wind-event period is a time of unknown risk. Stay inside until the radio advises it is safe to go out.

• *Do not touch wet electrical appliances.*

• *Stay away from fallen power lines, damaged buildings and particularly from bridges.*

• *Do not enter floodwater.*

• *Check for floodwater forcing sewage back up and into toilets, the kitchen and the laundry. If so (wearing rubber gloves) clean affected areas thoroughly using strong chlorine bleach or similar.*

MAJOR RISKS

The most serious risk is usually storm surge during and after the event. Wind driven waves surge, like mini tsunamis, well above normal sea levels — especially if coinciding with peak tides.

Flying debris and shattered glass pose serious dangers. While there is some risk inside a building, being outside is more dangerous. Big trees and even trucks may be thrown by winds of 300 km/h (185 miles/hour).

Wrecked fuel tanks may leak toxins that contaminate groundwater for years.

Electrical power may be cut resulting in banks and supermarkets closing for days. Water and fuel supplies may be unavailable due to no electricity to drive the pumps.

CHECKLIST: BASICS

Cyclonic events typically last three to four hours. Large ones can last longer. Shelter may be essential for up to eight hours as torrential rains often fall before, during and after such events.

• Move to a previously identified safe shelter in ample time. You are not usually allowed to take pets with you.

• Have your own, or access to, a strong cyclone resistant structure at least 10 metres (11 yards) above the highest known flood or storm surge level, and equipped to provide shelter for at least 24 hours.

• Set up access to basic food and water, plus the items listed later in this chapter. This enables you to live for at least two weeks without help from others.

• Have a battery or hand crank generator radio, at least two good torches, spare batteries for all, and matches and candles.

• *Have a map of the area extending a few hundred miles or kilometres out to sea so that you can plot wind tracks.*

WATER AND SEWAGE

Cyclones and their invariable torrential rain and floods result in two main health risks: sewage-contaminated water, and reptiles seeking shelter.

Floods overwhelm sewerage systems. Septic tanks fill and overflow. No matter how thirsty, *never* drink groundwater (it may kill you). It is risky even to wade through contaminated groundwater.

Have at least two week's stored water for drinking.

As long as storm water has not flowed into it, swimming pool water can be made safe to drink. Boil it for at least one minute, or add a little household chlorine (the taste disappears if left overnight).

To use rain water for drinking or cooking, when certain it is not contaminated by flood water, add any of the following for every 1000 litres (220 gallons).

Household chloride – 125 grams (4.4 ounces).

Granular swimming pool chlorine – 7 grams (0.25 ounces).

Liquid swimming pool chlorine – 40 grams (1.4 ounces).

Do not *exceed* such amounts.

It helps to have water for toilet flushing but mild laundry bleach (or Napisan) makes toilet waste safe.

When tap water is restored, clear the pipes by running water (with all taps open) for 15 minutes or so.

REPTILES

Snakes swept via floodwater seek refuge where they can. They will normally not attack unless threatened but snakes fleeing from floodwater will be stressed and can then be a very real danger.

If you encounter a snake outside, back slowly away. If it is inside, leave the room (with a door open ideally to the outside). Then cautiously, squirt *any* aerosol into the room. Snakes strongly dislike strange smells. They usually leave fast if subjected to unexpected odors.

Crocodiles and alligators may be swept into city and other streets. In the USA, this is most likely around the southern tip of Florida. It is also not uncommon in central and South America and the Caribbean.

Crocodiles are a risk across the top 1000 km (620 or so miles) of Australia. They can be swept by fast moving water to vast distances from their former territory.

These reptiles stalk submerged. In cloudy water it is impossible to tell if one is there until it strikes. When it does, it is at lightning speed over 10 metres (11 yards) or more.

Following a storm surge, sharks may be swept over 100 km (62 miles) inland. A big bull shark was seen in Queensland's capital (Brisbane), following extensive flooding.

MOSQUITOES

All mosquitoes are unwelcome. Some transmit potentially deadly diseases. They breed quickly in stagnant water following floods.

Where possible eliminate their breeding sites. Debris tends to contain breeding grounds, as also do old tires, tarpaulins, palm fronds, buckets, otherwise empty containers, boats, rain water gutters etc.

Wear loose light-colored clothing and socks. Use personal insect repellent if needed, plus home insect repellent. If feasible repair damaged insect screens.

PROTECTING YOUR HOME

Many of the more cyclone affected areas have associated cyclone building codes.

In the USA, the Miami-Dade Code requires exterior openings to be protected against wind-borne debris from hurricanes. An impact test is done on hurricane windows, hurricane window shutters and other hurricane impact-resistant products.

The Florida Building Code, International Building Code and Texas Department of Insurances codes are generally similar.

In Australia, new buildings in cyclone-prone areas must follow stringent standards.

Have nothing heavy, e.g. pot plants, barbecues, gas bottles etc. lying around outside; and no heavy trees close enough to the building to fall on it.

WINDOWS

Normal windows should be replaced by cyclone-rated frames with toughened glass, and cyclone-resistant window shutters added in cyclone prone areas.

The main Australian requirements are set out in the Standards AS/NZS 1170.2 – 2012 and AS 4040.3 – 1992. The USA's structural requirements are set out in: ASCE7-10.

Whilst toughened window glass is used in cyclone-prone areas, cyclone window shutters are more effective at preventing flying debris smashing through.

If shutters are not installed, secure mattresses against windows. Taping the glass also marginally assists: use the ultra strong 'duct' tape from hardware stores.

ROLLER DOORS

About 80% of residential hurricane damage starts with wind entry through garage doors.

Roller garage shutters may fail, allowing cyclonic winds to enter. A low pressure area develops at the rear, resulting in internal pressure causing partial or total collapse.

Fit cyclone-resistant roller shutters or strengthen the roller shutters temporally by a few strong steel beams or timber bracing.

ROOFS

Severe building damage, including collapse, is usually due to the roof (or whatever supports the roof) working loose. The cause is usually missing or corroded fastenings, and likewise roof battens etc. Another is termite rotted timber (termites are common in cyclonic areas).

Wind pulls the weakened timber around and tears it off. If the roof goes, the walls move and may collapse. Strengthen corrugated iron roofs by screwing them down at every rib for the first two/three rows, then at every second rib thereafter. Use the specially shaped washers and cyclone rated securing screws. Gable structures are also known to be at risk. They can be strengthened by triangular bracing.

Roof structures must be secured to the walls by metal straps, and/or the whole structure tied down by steel cables or solid steel bars. Diagonal bracing helps. Mechanical and structural engineers know how to do it.

Non-reinforced brick and concrete walls have inadequate strength. They may collapse – particularly in elevated or coastal areas.

Cyclonic rain is not just torrential. Once beyond 100 km/h (> 60 miles/hour) or so, it becomes almost horizontal. It will enter, via even tiny horizontal cracks, at high force. Test this using a fully-on hose with the jet nozzle held horizontally.

If you cannot make at least one place strong enough, it is best to find safe shelter elsewhere and in ample time.

RECREATIONAL VEHICLES

Do not even *think* of staying in any RV during a cyclone: none will withstand the wind forces and are liable to be thrown about and destroyed.

If a developing cyclone is within 300 km (about 185 miles) and you have no adequate shelter – now is the time to seek it. Do not panic or drive frantically, but don't delay as torrential rain may block exit roads. Drive to the closest town, ideally one with an RV park. Such parks are likely to have cyclone tie-down points. Place all loose items inside the RV. The staff will advise how to secure it and where to shelter.

CARS & 4WDS

Stay away from water — particularly the sea — trees, power lines, large signs etc. Pull the handbrake on hard and place the gear lever in 1st (or reverse if an automatic).

VITAL (CYCLONE) STUFF

Water is needed not just for the cyclone period but for about seven days after. Fill the bathtub with water (for later use).

Do not be overly concerned about having balanced meals etc for the first day or two. A few days on canned food makes no great odds.

ALSO NEEDED

Medical supplies: normal St Johns Ambulance basic kit, plus heart and other such pills.

Ample toilet paper: (supplies of almost anything may not be available for many days).

ENTERING DAMAGED BUILDINGS

Entry to any seriously damaged building is likely to be officially controlled: doing so without authority may seriously hinder or endanger rescue workers.

Less damaged buildings may be entered legally, but not necessarily safely if damaged by earthquake or fire; and likewise if they have been flooded.

Damaged but apparently stable buildings should not be entered until that entry has been formally approved by a structural engineer.

HOW TO ENTER - GENERALLY

Carry out the following actions *in the order listed* (page 44 explains why) before entering a building damaged by fire, flood or earthquake.

Do not strike any matches.

Turn off the gas at the main outside tap (see page 44).

Turn off the electricity at the main switch (see page 34 regarding solar arrays).

Check for fuel, water and sewage leaks.

Moving carefully, check for cracks or damage to chimneys, roof and walls.

Only then, with extreme caution, enter the building.

ENTERING DAMAGED BUILDINGS

AFTERSHOCKS

Earthquakes have before, main and aftershocks.
The main shock causes most damage. Aftershocks are
usually smaller and do less damage but 'usually' is not
'always'. The location, time and effects of aftershocks
cannot be predicted.

If possible do not work alone. If you are injured, it
may take a long time before assistance arrives.

In the event of an aftershock, leave the building,
checking carefully for falling masonry or other hazards.

SOLAR ARRAYS

Extreme caution must be taken with all premises that
have solar arrays. Many generate high and potentially
lethal voltage (300 volts or more).

Even with grid-connected systems that have grid
power disconnected, and/or grid power and inverter
switches turned off, the solar array and connecting
cables are likely to be lethally alive. And even more so
if the sun is clouded.

Turning off the associated inverter does not make the
array safe — it may still be producing high voltage.

Work on the system only if the modules are covered
against light. Or at night (moonlight is fine).

FIRE DAMAGED BUILDINGS

First turn off power at the main switch. Do not turn
it back on until a licensed electrician has checked all

wiring for damage. Insulation may be partially burned through. Appliances may have internal faults.

Wear sturdy boots (a common post-disaster injury is cut feet) and heavy gloves to protect against broken glass and smouldering debris. Ideally, wear disposable overalls, or ones that can be washed after each use.

Beware of smouldering material or hazardous substances including asbestos, toxic ash (copper chrome arsenic) from treated pine, garden chemicals and pool chlorine under debris.

Where feasible, keep debris wet, using a fine mist to avoid stirring up fine particles. Contact your local environmental protection authorities regarding safe and legal waste disposal.

DUST MASKS

Use a mask to safeguard against asbestos and fine ash. Copper chrome arsenic is of special concern (ash may contain these toxins). Even 1 to 2 grams (a tiny fraction of an ounce) is harmful.

There are three grades of dust masks. Those known as P1 and P2 are adequate for filtering fine particles including asbestos fibres.

The P2 masks filter out a slightly higher proportion of fine particles.

Grade P3 is required only in extreme conditions.

The US versions are brand-named *These.*

ENTERING DAMAGED BUILDINGS

Caution is required when entering a recently flooded building. Your *very first* action, whether power is available or not, is to turn off the electricity at the main switchboard. Then (if there is one) remove the main fuse. See also page 34 regarding solar arrays.

Residual current devices (RCDs) protect against electric shock but many older buildings lack them. Even where they exist, there is no guarantee they still work. (These devices are also known as ground fault circuit interrupters or appliance leakage current interrupters.)

If water has reached power point level do not switch the power back on without having the wiring checked by a licensed electrician or qualified electrical engineer.

Do not touch *any* electrical appliance before you have turned off the electricity as there is a risk of electric shock (from saturated appliances) if power suddenly returns.

Most appliances are useless after dirty water has penetrated. If battery powered, as long as they are (or were) not turned on, they can sometimes be saved by rinsing in totally clean water and *thoroughly* warm-air dried. Most, however, will need discarding.

SNAKES

Snakes, both poisonous and non-venomous, may seek shelter or have been washed in during the flood.

Whilst most are normally harmless unless disturbed, these may be highly stressed and dangerous to be near.

SANITATION

A dangerous and common post-disaster problem is seriously contaminated flood and storm water.

Check for any sewerage, and water supply damage. If possible avoid using toilets. Try to locate a plumber if you suspect damage to the sewerage lines.

Clean up the building and all in it, thoroughly disinfecting anything that water has contacted. Do this as described below.

Wear rubber gloves, gumboots and, if the water is deep, also watertight trousers.

If mains water is available, run all taps for 15 minutes to flush out any contamination that may have entered the system - then begin the clean up.

Remove any mud, then use bleach or disinfectant in really hot water to rinse everything clean.

Soak all utensils in a bleach solution for 10 minutes as bacteria penetrates deeply.

Because bacteria penetrates deeply, discard any wooden cutting board.

Air dry all items as it is almost impossible to keep cloths free from germs. Using them may cause cross contamination.

Mattresses are unlikely to be salvageable but sheets and blankets should be usable after thorough washing in really hot water. Sun-dry if possible – sunlight has a disinfecting action.

Carpets and rugs too can be disinfected by being left in the sun for a day or two. Expose both sides.

FOOD

Food that has come into contact with floodwater is almost certain to be contaminated. Throw it away.

Canned food may be salvageable if the cans are not dented or damaged. The cans should be soaked in bleach before opening.

PART TWO

The Needs

Electricity
Gas
Water
Toilets and hygiene
Communications

ELECTRICITY

The electricity grid supply shuts down in most disasters. This is not always because it is damaged but to safeguard people against electric shock from fallen power lines, and to enable electricians to work on the system. Power is usually restored within days, but sometimes weeks.

PORTABLE GENERATORS - A MAJOR RISK

Portable generators in such usage can be ultra-dangerous. They need to be close by, but doing so introduces serious health issues – and possibly death from carbon monoxide poisoning.

Do take this seriously. Following Cyclone Katrina (USA), over 50 people suffered severe carbon monoxide poisoning. Five died from it.

Australia's most severe cyclone ever (Cyclone Yasi in 2011) resulted in only *one* death – due to running a generator in a confined area.

That carbon monoxide rises. It enters through doors and windows. A generator needs to be at least 10 metres (11 yards) away from and *never below* a door, window or even air vent that's open.

Because of this, opening doors or windows is more likely to *cause*, not prevent, a potentially lethal build up.

Generators must *never* be used inside an apartment block nor any confined areas such as carports. Most of the post-Katrina accidents noted above were a *direct result* of using them in apartments.

The *only* safe place to use a portable generator is in the open and at a distance. This may not be possible, however, during floods. Further, a portable generator cannot be left in the open during major wind events (and may be stolen at any time).

Generators or other power sources absolutely ***must not be connected to any existing electrical wiring.*** By so doing, you feed power into the grid system that rescue people and electricity suppliers have shut down – and are repairing.

CARBON MONOXIDE SYMPTOMS

Carbon monoxide poisoning symptoms include dizziness, fatigue, headache, nausea, vomiting, and loss of consciousness. It is vital to get out of the affected area fast and, if possible, seek medical help. See page 117 regarding the risks.

SOLAR POWER

Solar may seem a good backup but many disasters are preceded, accompanied, and followed by heavy rain and long cloud cover that blocks the needed sun.

Smoke from (even remote) bush fires may slash solar input to close to zero for a week or more.

Earthquakes and volcanic eruptions do likewise via fine airborne dust.

Existing solar arrays are likely to be damaged and there is unlikely to be space for temporary solar arrays large enough to provide enough energy to be useful.

Solar, in post disaster situations is only worth pursuing by those wishing to experiment - but this is not the time to be doing so.

FUEL CELLS

Fuel cells are ideal for post-disaster situations. They are virtually emission-free — safe even to run 24/7 in a closed room, and are as quiet as a small computer.

The commonly used EFOY range produce 12 volts direct current at 900 to 2200 watt-hours/day. The smallest will power a few LED or halogen lights, small water pumps, a small LED television and a small chest fridge.

You can use small grid voltage appliances via an inverter, but fuel cells are not suitable for high power devices (such as electric kettles or microwave ovens).

Fuel cells need a small car battery for short term loads larger than they can instantly support. A large battery is not needed because fuel cells supply smaller loads indefinitely — as long as fuel is available.

Most fuel cells run from non-refillable methanol cartridges (but are currently only generally available from specialised suppliers).

A couple of 10 litre (2.2 gallons) methanol cartridges are likely to supply enough energy for two to four weeks. It pays to have a spare cartridge (or two).

The downside of fuel cells is that, currently about A$5000 or so upward, they are not cheap, nor are those methanol cartridges.

LIGHTING

Light emitting diode (LED) lights are ideal for post-disaster situations. One or two 5 watt units will provide all the light you are likely to need. The smaller dry battery powered LEDs are preferable. Use ultra long life lithium-ion or nickel cadmium batteries — not the cheap batteries often supplied with the units.

The headband LED units (from camping stores) are handy when you are moving about outside, and for cooking and reading.

Also have one really good LED torch for each person, plus at least one spare.

Many people suggest using candles, but whilst candlelight is fine for those under 40, older people need a lot more light. Candles are also an unnecessary fire risk.

By all means keep a few candles in reserve, but even the smallest LEDs produce far more light than half a dozen candles at once. Battery powered LEDs typically run for a month or more (even if used every night).

GAS

Gas, for home and industrial use, originally contained toxic carbon monoxide. It was deadly if inhaled.

Natural gas (and LP gas) is less toxic, but can kill if they replace the air you breathe. Their major risk sources are fire or explosion due to damaged pipes — or from carbon monoxide generated by burning such gas in non-flued appliances.

Natural gas has no odor, so bad smelling gas is added to warn of leaks. In immediate post-disaster situations, even if you cannot smell gas, turn off all gas appliances and then turn the gas off at the main tap.

Only then, turn off the electricity. Never turn the electricity off first — because doing so can create a gas-igniting spark and instant explosion.

Ensure again that all gas appliances are turned off. Then turn the gas back on and check each appliance for leaks by coating liberally with dish-washing liquid or any oil and checking for gas bubbles.

This is especially necessary after even minor earthquakes.

LP GAS
Unless designed (or modified) to run on LP gas, domestic gas appliances may only be run from natural gas. Because of this include a small portable LP gas

stove ring and a few gas canisters in your back-up equipment and supplies. Use only for cooking — and in a well ventilated room.

It is often suggested to use a portable barbecue. This is not a good idea. They are dangerous if hurled around outside in cyclones. They are also prone to be stolen.

The risk of carbon monoxide poisoning is far too high to even *think* about using a portable barbecue indoors. See page 110 regarding this.

WATER

People survive without food for many weeks. Without water they only last a few days. Water transports glucose, oxygen and fat to muscles, and waste products away from muscles (as urine).

Water also aids digestion, lubricates joints, organs and tissues and maintains the body's working temperature.

Most people need about two litres (half a gallon) of water a day. Beer, wine and soft drinks do not substitute.

Signs of dehydration include disorientation, dark urine, fatigue, headache, irritability, nausea, muscle cramps and weakness, but not necessarily thirst. You may feel hungry when you are dehydrated but food is far less important than water.

WATER RISKS

Major wind events and floods often overwhelm public sewerage systems and cause water pumping infrastructure to fail. Raw sewage may be forced back into your home.

Earthquakes cause anything semi-liquid, such as saturated soil, to sink otherwise soil-held sediment.

Such so-called liquefaction can trap people, and cause cars and even whole buildings to partially sink. It also forces sewage and other contaminated water to rise (dangerously) to ground level.

Water supplies may be cut off for days. Ground and tap water is likely to be too dangerous to drink.

Do not even *wade* through groundwater unless wearing waterproof gear.

STORING WATER

Store enough pure drinking water to last two weeks. The simplest and safest way is large sealed plastic water-filled containers pre-purchased from supermarkets.

Such water may acquire a slight plastic taste over time but will not develop algae or bacteria unless opened (it is then safe to drink for about two weeks).

Obtain and store *commercially bottled* water now. Don't wait until a disaster situation is imminent: it sells out quickly due to panic buying.

Expiry dates may be legally required but it is generally held that *commercially bottled* water is safe indefinitely. This is *not* true of self-bottled tap water.

STERILISING

As long as you are 100% certain that no storm water has flowed into the swimming pool, if boiled for at least one minute, that water is safe to drink

It may also (or alternatively) be treated with a quarter to half a teaspoon of household chloride, or 10 drops of swimming pool chlorine per 11 litres (2.2 gallons) of water.

If treated as above you can use pool water for a time without risk to health. The chlorine taste is reduced if the treated water is left to stand overnight.

FILTERING

An alternative is commercial filtering. This requires a 10 micron filter followed by a 1 micron filter (0.5 micron is better but less readily available). Filter elements must be changed yearly to remain effective and ensure they remove cryptosporidium and giardia.

As neither mains pressure water, nor electricity may be available have a *small* hand pumped filter for this purpose. (The larger ones need mains pressure water).

Once mains water returns, run all taps fully open for 15 minutes to clear contaminated water from the public supply and from your own water system.

SOLAR DISTILLATION

Solar stills made from plastic sheets are *marginally* effective. To make one, cover a receptacle or large hole in the ground with clear plastic. Then place a small weight in the centre to form an upside down plastic cone. Moisture distills on the underside of the plastic and drips into a receptacle placed under that weight.

A small tube from the receptacle to the *outside* of the still enables you to readily access the distilled water. Output is increased by placing plant material or anything vaguely damp inside the still.

   Solar distillation can save you but the following is far better as long as *some* water is available - even if contaminated.

SODIS

SODIS, an acronym for solar water disinfection, is a UNICEF and World Health Organisation officially approved way of disinfecting water using sunlight (or even just daylight) alone.

   It works by using the sun's long-wave ultra violet rays to kill viruses, bacteria and parasites such as giardia and cryptosporidium. Heat is not required. The process works equally well in low air and water temperatures.

USING SODIS

Find a glass drinking water bottle or a clear standard PET plastic bottle (one with a bluish tint is fine). Do not use large containers because UV radiation cannot adequately penetrate them.

   Most plastic bottles that have held anything drinkable are PET and labelled as such. If unsure try burning one. PET burns quickly and readily (when in a flame) and smells sweetish. Most other plastics do not burn readily. If they do, they have a pungent smell.

   Water to be treated must absorb light. To test, place a filled bottle on a newspaper. If you can read the headlines the water is usable. If you can't, filter the water through a cloth until it has sufficiently cleared.

Wash the bottle/s with clean water and disinfectant. Fill with the contaminated water and leave it in the sun, ideally on something reflective such as light colored corrugated iron, or paper.

Six hours exposure is enough if the sky is no more than half clouded. If the sky is more heavily clouded, expose the bottle for two days in a row.

SODIS does not work when it's raining — but you then have drinkable rain water anyway!

Drink from a personal SODIS bottle or from a receptacle known to be germ-free. The water may not taste pleasant: the SODIS method cannot change that but the treated water is germ-free.

Be aware SODIS only kills *germs*. It cannot neutralise contaminants such as poisons or fertilisers.

Storing SODIS treated water in a cool dark place ensures it will remain safe to drink for as long as necessary. Algae may form but is harmless.

WATER TANKS

Many people have rain water tanks primarily for irrigation. If you use an existing tank for emergency use, unless almost full, it may be blown over in a cyclone, washed away by storm surge, or contaminated by flood and storm water.

Such tanks need to be partially buried and kept close to full (or they may float out), and ideally located on high ground.

Rain water is legal for washing and toilet flushing but some jurisdictions forbid its use for drinking.

You are unlikely to get into trouble for drinking it in an emergency situation but before doing so run it through the type of filter described on page 48.

It's also feasible to collect and store rain water — but recycle it from time to time. If you intend to drink it, filter it first.

Drinking water apart, the smallest amount of water worth storing is around 3000 litres (660 gallons). A tank this size fits in most gardens. It will supply enough water for two or three people for 14 or so days.

Have a gravity fed tap a little above the bottom of the tank. Avoid electric pumps as grid power is unlikely to be available.

Solar cannot be relied upon for such a vital need and it's not worth having a generator for such minor draw.

If pumping water over distance use a hand operated unit made for the RV market. Ensure it can suck up water from whatever level you need.

KEEPING CLEAN

Unless you have adequate assured water, forego having showers except during heavy rain. Instead, use a flannel and soap for washing, and perhaps a 10 second rinse using a fine spray.

WATER

HOT WATER WARNING

Some books recommend heating water within black poly pipe placed in the sun. The pipe outlet is left open and held vertically to retain the water until used.

Such books do warn that outlet must *never* be closed off — as the pressure of the heated water may cause the pipe to expand and burst. It would also be ejected at high pressure when accessed.

Whilst heating water this way is simple and effective it introduces a real risk of severe scalding — the water may reach boiling point and be ejected at high pressure.

In the author's opinion it is potentially too risky to use, particularly if young children are involved.

This book thus warns (strongly) against using it.

A far safer way is to leave a large shallow vessel of water in the sun. Heat insulate it underneath by towels, blankets or newspaper etc.

TOILETS & HYGIENE

Following the 2011 Christchurch (New Zealand) earthquake, the entire city was left without water, or usable home and office toilets, for a week or more.

Shortly after, the Tōhoku earthquake and tsunami, plus subsequent nuclear reactor disaster left part of Japan in a similar situation.

Even where toilets and flushing water are available, raw sewage may back up and even overflow, rendering toilets unusable. If this happens there is likely to be no choice but to arrange a temporary alternative.

OPTIONS

If the toilet bowl is clear, but lacks water for flushing, line it with one (disposable) garbage bag inside another. Use heavy duty garbage bags made for garden waste.

If that is not an option, use a steel bucket, with a strong handle, and some form of tight fitting lid, or an empty oil drum. If necessary, improvise a seat out of heavy cardboard or timber. Or take a seat from an existing non-usable toilet. Line with two garbage bags as described above.

A better solution (and not that costly) is to keep a portable toilet for this possible occasion.

SANITISING

Specialized chemicals are available to break down sewage, neutralise its odour at least partially, and render it safe. Some of these chemicals, however, are based on formaldehyde, a substance long recognized as very bad for the environment.

Recreational vehicle users, and people living in remote areas, often use baby nappy treatment bleaches available from supermarkets for only a few dollars.

They are less environmentally harmful and work just as well as the specialised products. They work by releasing oxygen that breaks down the bacteria.

Use about 10 tablespoons per full sized bucket of water. Add it only *after* a couple of 'number twos' and then whenever the load is added to. (Apologies to the squeamish - but it needs something 'to work on'.)

The treated sewage is reasonably bacteria-free. It can be buried deeply in the garden but kept well away from any water sources and anything growing that you may eat. It also needs to well covered by soil. Add a few rocks (to stop dogs digging it up).

Empty the container before it gets too full. The treatment will have made it less dangerous but cleaning up any spill is not a fun thing to do.

Water for washing always comes second to water for drinking and cooking.

Baby wipes provide rapid hassle free body cleansing. Use them also to sanitise your hands after handling raw meat and wiping things clean. If the package is unopened they remain usable for two to three years.

Despite labels to the contrary do not flush these down a toilet. They can and do block entire sewerage systems.

Use disposable toothbrushes. If you also use dental floss they work well enough without water.

Store tampons and sanitary napkins if you are likely to need them. They can be kept for a long time and are often overlooked items that may be needed post-disaster.

COMMUNICATIONS

Access to warnings and updates is vital, particularly for fires and tsunamis. Also necessary and comforting is knowing what's happening locally.

This can be via phones, laptop computers, radio and TV but, as electrical power is likely to have failed or been cut off, have at least one system that runs from replaceable dry batteries – and/or have a short term back up power supply (pages 40-43).

PHONES

Whilst home phones may fail in disaster situations, home phone numbers provide exact locations for emergency services. The handset must be a basic one powered via the telephone exchange — not grid powered — as that may fail or shut off.

A mobile phone is handy but may not be usable if the associated antenna towers are damaged. Further, mobile phone services can only handle a limited number of calls at a time.

SATELLITE TELEPHONE

The most reliable communication is via satellite phone. They are costly to buy and use but enable you to call any telephone number world-wide. They will not be affected by anything happening locally, but must be able to 'see' the satellite. An outside antenna is usually needed if used inside a building.

COMMUNICATIONS

CB RADIO

The citizen's band (CB) radio service is known in some countries as the General Radio Service and the Personal Radio Service. Most services have 40 channels (but a few, including Australia) have 80, with one channel, usually '9', reserved for emergency use, and monitored during disasters.

Reception distance varies with terrain but is typically reliable over about three miles (five kilometres).

AM/RADIO

An AM/FM battery or hand crank powered radio is a must. During cyclones, fires or earthquakes, ongoing radio reports inform of their locality, severity and likely duration. They also provide local advice.

Check for reception in your planned secure area. If poor, a temporary antenna assists. External antennas are better but few withstand even low level cyclones.

TELEVISION

Whilst less vital than radio, TV can be comforting and advises of emergency services work and extent of damage.

Battery and wind up generator powered TVs are available. An iPad etc can receive TV via a plug-in antenna. As with radio, an external antenna is likely to be blown out of alignment or torn away.

COMMUNICATIONS

INTERNET

The internet may lack up to date and relevant emergency information but assists if coupled with that from radio.

The 3G and 4G (and 5G in late 2018) communication services offer a reasonably reliable form of speech and Internet communications in disaster situations.

Transmission frequencies for various devices (phones, notebooks, lap tops or modems) vary from country to country.

Multi-band devices are available. They require a SIM card specific to the country and service provider.

A tiny antenna/modem built into a USB will generally suffice but fringe areas require specialised units.

PART THREE

Keeping well

Keeping warm or cool
Medical

KEEPING WARM OR COOL

Keeping warm post-disaster can be a major problem especially as electricity and gas are likely to be cut off.

Serious cold, especially accompanied by wind chill, kills fast. It was a major cause of death in the first days after the 2011 Japanese earthquake and tsunami. Keeping warm, therefore, can be a vital priority.

The simplest way is to conserve the body's existing heat. This is readily done by blankets buyable from army disposal stores for a few dollars each.

If no blankets are available, you can keep surprisingly warm by forming a sleeping bag from two heavy duty plastic garbage bags. Insert one inside the other, carefully lining the space between with old newspapers. Or use newspaper to line your clothing.

This is so cheap, simple and effective that there is little point in seeking *costly* alternatives.

COOLING DOWN

Keeping cool matters less. It is rarely an issue during and for some time after a cyclone: the usually heavy rain drops temperatures to a comfortable level. The temperature may rise again a few days after a cyclone.

Being too hot is uncomfortable and debilitating (particularly if washing water is scarce). It does not, however, present the serious risks of extreme cold.

KEEPING WARM OR COOL

Cooling off entails avoiding as much heat as possible, then allowing and assisting some body heat to escape.

It is harder than keeping warm, and there are limits to what is possible, but a lot can still be done.

WHAT TO WEAR TO KEEP COOL

Wear the absolute minimum culturally practicable: ideally loose fitting shorts and tops of natural-fabrics such as cotton or linen. Bare feet are best providing there is no risk of injury or infection – but do so only within your safe area.

Minimise heat build-up by having windows open at night (if safe and mosquito-free) but closed if its hotter outside than in. And vice versa. If there's little difference (but breezy), open windows selectively.

Even small fans help but to have them work during power cuts requires some form of stored or self-generated power.

Never fan yourself: it only *seems* cooler: body heat generated by the energy so used exceeds the heat lost.

Whilst only possible with non-contaminated water, spraying yourself and/or your clothes with cool water works well, particularly in a breeze. So does pouring cold water over your wrists (and inside elbows) and also wearing cold wet socks.

To keep cool in bed sleep with a dry towel beneath you and a wrung out wet sheet on top.

MEDICAL

This section provides medical information only.

It is not medical advice, nor intended as a substitute for medical advice, diagnosis or treatment.

The content is sourced from public domain professional medical material (from the USA, Canada and Australia) *for information only.*

It is provided without any representation or warranty that it is complete, correct or current.

It is absolutely not intended to be used as an alternative to professional medical *advice* of any kind.

Reliance on any medical information in this book is thus solely at your own risk. Unless *totally impossible,* seek the advice of a physician or other qualified medical health care provider.

Never disregard professional medical advice or delay seeking it because of something you have read here.

A truly excellent source of medical information in emergency situations is the Hesperian organisation's free book: *Where There Is No Doctor.*

It is now produced in over 80 different languages (including 13 that are African). See page 114 for details of how to obtain it. If feasible, download it NOW.

WATER IS TOTALLY VITAL

Humans may survive for months without food, but for only a few days without water.

General medical opinion is that it is vital to drink at least two litres (half a US gallon) of water a day. Avoid most other drinks (particularly alcohol).

Medical issues common to most post-disaster situations are stomach upsets that result in at least minor dehydration.

DIARRHOEA

The purpose and function of diarrhoea is to rid the body of stuff better not eaten – and discharged accordingly. It usually fixes itself in two/three days.

Diarrhoea is the most likely medical condition to occur after flooding and/or storm surge. It is essential to pay scrupulous attention to avoid contacting, and especially drinking, contaminated water.

Doctors stress that care is needed to avoid dehydration as diarrhoea hinders absorption of nutrients and water.

EMERGENCY TREATMENTS

All of the following require emergency treatment. Every possible effort must be made to obtain it.

Abdominal pain that persists after a bowel movement.

Vomiting or diarrhoea in a baby under three months old (urgent).

Older babies who have been vomiting for more than 12 hours.

Bloody, black or oily looking stools.

Diarrhoea extending beyond five days.

Fever (accompanied by diarrhoea) over 38.3° C (101° F) in adults or 38° C (100.4° F) in children.

Symptoms of dehydration (e.g. dizziness, weakness or muscle cramps).

COMMON INFECTIONS

The most common infection is by bacteria, mainly via campylobacter, salmonella, E.coli 0157:H7 and the group known as calicivirus.

CAMPYLOBACTER

Campylobacter is in the intestines of healthy as well as sick birds. Most raw chicken carries it. It may be transmitted via insufficiently cooked poultry, unpasteurised milk or contaminated water.

Its incubation period is one to seven days. Symptoms include fever as high as 40° C (104° F), headache, abdominal cramps and diarrhoea.

Reference sources state that doctors typically advise that medical treatment is not *normally* required for campylobacter except in the event of high fever, bloody diarrhoea, more than eight bowel movements a day and worsening symptoms – or those that persist for longer than a week.

Further, doctors state that medical treatment is essential if the sufferer is pregnant, has HIV, AIDS or other immuno-compromised conditions.

SALMONELLA

Salmonella symptoms are as for campylobacter. It is common in bird, mammal and human intestines, contaminated water, soil, animal faeces (especially those of reptiles such as turtles and lizards) and in raw eggs and uncooked meat. It is typically transmitted to and from food preparation surfaces by inadequate hygiene.

Doctors typically state that symptoms normally decrease without medical treatment after a couple of days - and that sufferers (particularly children) need lots of liquid *and medical attention* if they develop abdominal pain, high fever, bloody faeces or do not recover after a few days.

Doctors also typically state young babies *need medical attention* if bowel movement is looser or more frequent than usual or if there is vomiting; and that people with poor immune systems may need an antibiotic.

E.COLI 0157:H7

E.coli 0157:H7 causes severe and bloody diarrhoea and painful abdominal cramps, but rarely fever. It is related to cattle and other animals and usually results from consuming food or water contaminated by cow faeces, eating raw or under-cooked ground beef. Under-cooked hamburgers are a serious risk as is raw milk.

Symptoms usually appear after three to five days (but may vary from one to 10 days). As with most such issues symptoms include nausea, vomiting, stomach cramps, and diarrhoea (often bloody). There may also be mild fever: 37.7^0 C to 38.3^0 C (100^0 F to 101^0 F).

Medical advice is that, if possible, E.coli 0157:H7 infection should be treated by a doctor, especially for children or elderly people. Typical reports are that most people recover without medication in five to seven days but about 10% do not. *Medical advice is necessary if symptoms last longer than seven to eight days.*

GASTROENTERITIS

Gastroenteritis (a calcivirus) too causes vomiting and some diarrhoea but usually lasts for only two or so days. Its only source is via faeces from those infected and is primarily spread from person to person.

It is particularly associated with oyster catchers and kitchen workers with inadequate hygiene habits. It is also spread by raw sewage found in storm or floodwater.

Here too, symptoms include nausea, vomiting, diarrhoea, abdominal cramps and headaches – but typically only low level fever.

Not all victims develop all of the symptoms. Some may just develop a mild fever of about 37.7^0 C (99.8^0 F). Doctors say that most symptoms resolve in two to five days.

Medical advice is typically that gastroenteritis does not usually require medical treatment but needs ample liquids to prevent dehydration. Those recommended include water and electrolyte replenishing sports drinks, but not fruit juices nor milk.

Such advice is also that medical attention is needed if symptoms last longer than about five days, or increase in severity; or if there is a fever of $38.3^0 C$ ($101^0 F$ or higher. Or if bloody diarrhoea, dehydration, constant abdominal pain, or other symptoms develop.

Doctors advise that if young babies are part of the group, to breast feed them if at all possible. If using formula, it is *vital* to use bottled water, or to boil it for at least one minute.

DEHYDRATION

The symptoms of dehydration include any or all of the following: decreased or no urine, dry mouth, dry mucus membrane, dry skin, feelings of weakness, an inability to produce tears, light-headedness and low blood pressure.

Dehydration is a serious risk. Unless medical advice forbids, drink lots of clear fluid and/or sports drinks. Avoid alcohol or caffeine. Milk usually prolongs diarrhoea but (in very mild cases) can assist by providing nutrients.

Active cultures such as probiotic yogurt (if available) ease and shorten symptoms of some types of diarrhoea. So do bananas, rice, apples or apple sauce and dry toast.

GENERAL BLEEDING

Medical advice here is to keep the victim lying down, then use a clean cloth (ideally a sterile dressing) to cover the wound using firm direct pressure.

Do not attempt to remove any lodged object. Instead seek professional assistance as soon as possible. If the injured area has no fracture, most medical advice is to ease the body such that the wound is higher than the heart and to keep the victim warm and, particularly for major wounds and other injuries, check regularly for signs of shock.

The main symptom (of shock) is low blood pressure. It may typically be accompanied by any or all of the following: cold and clammy skin, dizziness, fast and weak pulse, general weakness and rapid, shallow breathing.

There may also be staring eyes, anxiety, bluish lips and finger tips, chest pain, confusion and sweating.

SHOCK

Medically, 'shock' does not imply mental reaction. It is a serious condition in which body tissues lack sufficient oxygen and nutrients for cells to function.

Initially, the body tries to maintain normal blood pressure by moving fluids from cells to the bloodstream. In doing so, breathing and pulse rates progressively speed up.

Blood pressure then starts to fall and compensation mechanisms cease to act. This leads to bodily responses such as organ failure and loss of consciousness.

Paramedics advise that, if you come upon a person in apparent shock ensure they are in a warm safe place, and **immediately** seek emergency medical help.

CLEANING AND BANDAGING

Sterilise your hands then clean the wound with soap and water. Dab dry with clean cloth. Apply antibiotic and cover with a bandage or sterile gauze.

EYE INJURIES

The general published advice is never to attempt to remove an object embedded in an eye. It is to cover *both* eyes with sterile dressings and seek emergency help. (covering both reduces movement of the *damaged* eye).

If injured by any chemical, the general recommendation is to move the victim's head such that the affected eye is lowest and flush with clean water at room temperature for 10 to 15 minutes. Also, if the injured party is a contact lens wearer, to remove the lens after flushing and then flush for a further minute or two.

BURNS: THERMAL

Typical published advice is that professional attention is not necessarily needed if the skin is red and possibly swollen, but not blistered.

If the skin is as above *but blistered*, seek professional attention (if available). In both of the above situations medical advice is to initially submerge the burn area in cool water for at least 20 minutes.

When pain stops cover with cool wet cloths.

Do not break any blisters. If pain continues, obtain professional attention as soon as possible meanwhile covering with a sterile dressing. Do not apply any ointment.

Do not attempt to remove clothing sticking to a burned area. Cover all with sterile gauze and seek medical assistance urgently. Such burns are emergencies. They must be treated as such.

BURNS: CHEMICAL
Remove all contaminated clothing, watch and jewellery. Flush with cool uncontaminated running water for at least 15 minutes.

Watch for shock symptoms and seek medical assistance if at all feasible. If the eye is chemically contaminated or burned, flush for as long as feasible. Seek emergency medical aid.

UNCONSCIOUSNESS
CPR (cardiopulmonary resuscitation) courses advise to check the airway for obstruction, breathing and circulation (pulse) and then commence CPR.

They further advise that, if the airway is not obstructed, breathing and circulation are working and there's no indication of spinal damage, to roll the victim onto the side (chin toward ground) and cover with a blanket.

CARDIO PULMONARY RESUSCITATION (CPR)
This potentionally life-saving technique cannot be safely learned from books, not least as it is progressively updated in light of global experience.

CPR courses are available almost globally, particularly from St John's Ambulance. It is strongly advised that all readers do one of the (usually one or two day) courses at least every second year.

If CPR training is totally unfeasible, Googling CPR (and the current year) will uncover a number of excellently presented current videos. See also page 114.

POISONING

Call emergency services if at all possible.

Stock your medicine cupboard with a medicine that causes people to vomit *but use it ONLY if advised to do so by a professional medic.* It must not be administered to anyone who is drowsy, unconscious or convulsing or has ingested strychnine, corrosives, petroleum distillates, Lysol, solvents, thinners, detergents or anti-emetic drugs.

Keep the emetic under tight security at all times. It is a potentially dangerous substance.

If poison is on the skin, the general advice is rinse with plain water for 15 minutes, then soap and water.

For poison in an eye, such advice is to flush with lukewarm water for 15 minutes. Then, as with eye injuries, *seek medical attention as soon as possible.*

SNAKE BITE

Do not wash off the venom. It doesn't seep through the skin nor into the wound and is vital for identification.

Using a wide elastic bandage, bind the limb firmly towards the extremity and then back toward the trunk.

Immobilise the limb by strapping it to anything available (such as a broomstick). Record the time (doctors need it) *and seek urgent medical attention*.

FIRST AID SUPPLIES

Be cautious when choosing. Some suppliers claim their kits are designed for 'disaster situations'.

These, or promotional material, may carry statements such as 'The mission of the Occupational Safety and Health Administration (OSHA) is to save lives, prevent injuries and protect the health of America's workers'.

This does not imply association or endorsement. OSHA endorses products *only* for typical and specialised workplaces - not disaster situations.

The International Standard (ISO 7010) requires that first aid kits are identified by a green and white symbol (left). It does *not* define their contents.

The red cross on a white background (centre) is often used by International Red Cross, and as a Johnson and Johnson trade mark (for medicinal and surgical plasters) and also by the military. Here too, it only advises there is a first aid connection. It does *not* define content.

The Star of Life (above right) is associated with

emergency medical services, and to indicate that the service using it offers (or is) an appropriate point of care.

Few standard first aid kits contain the range of medical supplies and equipment that may be needed in a disaster.

Instead, assemble your own, using quantities to suit specific needs.

FIRST AID KIT SUPPLIES

Antiseptic sprays, baby wipes and swabs.

Bandages: 50 mm, 75 mm and 100 mm (2", 3" and 4").

Bandages (elastic) for providing support and/or pressure.

Bandages (adhesive).

Bandages (triangular – about one yard (metre) for slings, splints, tourniquets etc.

Bandaids (various).

Blood pressure monitor (if applicable).

Catheter tip syringe for cleaning wounds.

Disinfectant (mild).

Dressing (burn – sterile pads containing a cooling gel).

Saline (for cleaning wounds and washing out eyes).

Foil blanket ('space' blanket) also doubles a sun reflector.

Eye pads (sterile).

Eye protection – disposable goggles.

Face masks (disposable). See References on page 35.

MEDICAL

First aid notes or book plus note pads and pencils.

Gauze swabs (sterile).

Gloves (disposable), and others of (kitchen strength).

Magic markers.

Magnifying glass – for seeing small splinters and objects in eye.

Plastic bags (various sizes).

Plastic sheet, about 3 metres by 10 metres (or yards).

Pocket knife with spring blade or Swiss Army knife.

Pocket mask plus face mask for artificial respiration.

Prescription medicine (for two weeks).

Prescription glasses and sunglasses.

Safety pins.

Saline (for cleaning wounds and washing eyes).

Scissors (blunt, sharp, stainless steel).

Soap (needed to clean minor wounds after bleeding has ceased).

Splinter forceps (stainless steel).

Sunscreen.

Tissues (several packets).

Thermometer.

Trauma shears (double as scissors but are more effective for cutting away clothing etc.).

Tweezers (sterilised before and after use).

MEDICAL

MEDICATION

Commercial first aid kits contain basic medications. It is advisable to put together personalised kits. That included should first be discussed with a doctor.

The following includes only examples of what *may* be included. (Cycle prescription medicines before they expire: see page 76.)

Antibiotic cream.

Anti-coagulant such as aspirin for minor chest pains.

Anti-diarrhoea medication – the risk of diarrhoea is high in *flood and storm water situations.*

Antihistamines for allergic reactions.

Anti-inflammatory (e.g. Ibuprofen, Naproxen).

Pain killers such as codeine and paracetamol.

Prescription medications etc – page 76.

TOPICAL MEDICATIONS

(Topical medications are those applied directly to the skin.)

Adhesion aid – tincture of benzoin protects the skin and aids the adhesion of butterfly strips or adhesive bandages.

Aloe vera gel for minor burns (including sunburn), itching, dry skin. It may also be used as an antibiotic gel to keep wounds moist and prevent coverings from sticking.

Anti-itching preparations. They include hydrocortisone

cream and antihistamines containing diphenhydramine.

Antiseptic fluids and ointments include benzalkonium chloride, neomycin, iodine and bacitracin zinc.

Burn gels are water-based gels that cool. They may include a mild anaesthetic.

MEDICAL RECORDS
Knowing a patient's medical history and/or required medication is vital. Delay in obtaining it may be critical for survival, especially if there are language problems.

See the example on page 115. Copy and complete one for each person likely to be involved. Have several copies readily available including one that you keep with you at all times.

PRESCRIPTION MEDICINES
For those reliant on prescription medicines and drugs, many doctors will provide four or five month's repeats. Pharmacists will usually supply you with that amount if you explain why they are required.

Manufacturers in this area set an expiration date that denotes the final day that the full potency and safety of a medication can be assured. If unopened, this is typically one to five years. Once opened, however, that expiration date can no longer be relied upon.

For legal and liability reasons, manufacturers will not make recommendations about the stability of drugs past the original expiration date.

PART FOUR

Keeping Fed

Eating & drinking
Cooking

EATING & DRINKING

Ample water is vital. Without it, most people die within five to seven days. Lack of food, however, even for two weeks, is uncomfortable but, except for diabetics and a few others apart, it is not a major health risk. Despite this, it makes sense to store enough basic foodstuffs to last for two weeks.

This food needs to be canned and/or freeze dried. Neither is 100% healthy, but if you include canned or (ideally freeze dried) fruit and vegetables you are likely to be fine over that time.

(As noted elsewhere in this book, babies are best breast fed if possible. If using formula, use bottled water, or boil water for at least one minute.)

CANNED FOOD

Canned food is safe to eat for many years. Its taste gradually deteriorates but is acceptable for four or five years (or even longer if kept cool).

Cans are rugged and vermin proof, but need to be kept dry. If not, they rust, rendering the contents inedible.

Canned meat varies from acceptable to all but inedible, but canned chicken is usually good.

There are canned beans of all varieties. As the cooking suggestions (pages 83-89) show, green beans, peas, carrots and mushrooms plus rice or pasta, are the basis of excellent and nutritious meals. Canned pears, peaches and various berries provide vitamin C and other vital nutrients.

Canned food is often usefully high in calories, and also contains vital water.

Discard cans that bulge, are damaged or partly used.

Cans may have labels destroyed or made unreadable by contact with water. To avoid having some very strange meals, set up a totally foolproof system to identify their contents – using indelible marking pens.

CONTAMINATION

Cans that may have been in contact with flood or storm water must be decontaminated before opening. Note down the contents, then remove the labels. Wash the cans thoroughly in clean water and then soak them in about 50 millilitres (two ounces) of unscented bleach per two litres (half a US gallon) of water.

Soak for two minutes and rinse the cans immediately. (This works *only* with steel cans. It is almost impossible to sterilise aluminum cans.) Then mark the contents on the can/s with an indelible pen.

DEHYDRATED & FREEZE DRIED FOOD

Dehydrated and (the preferred) freeze dried food has a shorter (but adequate) shelf life than canned food.

The lower the storage temperature and the drier the food, the longer it lasts. Such food is usually cheaper, more compact, light to carry, easy to open and prepare, but requires water to restore its use.

Do not try to dry food yourself. It is hard to remove sufficient moisture – limiting shelf life. The food must be insect free and the packaging totally vermin proof.

READY TO EAT MEALS

Developed for the space program and then used by the US military, meals ready to eat (MRE) are commonly bought by backpackers. Whilst handy, they are costly, high in carbohydrates and lack essential roughage.

Many companies produce food packs with varying claims of their content. Most require heating in boiling water but some have a flame-less (chemical) heating pouch. The products *can* be eaten cold as they are fully pre-cooked.

Most have a claimed shelf life of three to five years but some claims are of 'up to 40 years'.

LONG STORED FOOD

Do not consider storing frozen food for any post-disaster situation as electrical power is very likely to be lost. Apart from the risk, it is pointless to generate electricity to freeze food: alternatives need no power.

Consume any frozen food first. If unopened in a closed freezer it should stay safe for three/four days without power.

Oils keep for at least a year. Sugar, salt and honey virtually last forever if sealed and watertight. Pasta, rice and dried beans have long shelf lives in cool climates.

Dried herbs in glass jars kept in a dry dark place have a long shelf life but lose flavour after a few years.

Parmesan and most other hard cheeses have long shelf lives, but as with flour, it is best to use and renew them every few months.

If using cans, buy only sizes that will be consumed in one meal. Unless refrigeration being (improbably) available do not use left-over canned food.

Some cans with ring pulls are impossible to open without a can opener. Don't use a cheap one. They have ultra sharp edges that can severely damage your fingers and hands.

Have at least one really good *manual* can opener. Buy one now - before it is needed.

Discard screw cap containers that have been in contact with flood or storm water. They are impossible to safely disinfect.

ALCOHOL
Never drink alcohol immediately post-disaster. Unexpectedly dangerous events (such as earthquake aftershocks) may occur: you cannot afford to be even mildly alcohol-affected.

Certainly keep a bottle or two for later celebration, but reserve it until officially advised danger has passed.

FOOD PREPARATION SAFETY

During and often after any large scale disaster, medical aid may be unobtainable even for total emergencies. Keeping well is thus vital.

The major and *very* real risk in many disaster events is sewage-contaminated water (pages 46-52).

Physical injury apart, the next highest risk is food-borne illness. Take this *very* seriously, particularly if babies or young children are present. Ensure everyone in your household *always* does the following.

Wash hands before handling any food.

Never cut up raw chicken on wooden chopping boards. Contaminants get into their crevices. It's then virtually impossible to render them safe to use.

Discard or (ideally) burn wooden chopping boards if there is the slightest chance they have been wetted by flood or storm water.

Clean and then sanitize all food preparation surfaces as follows.

Wash with hot clean water with soap or disinfectant washing up liquid. Then immerse or flood using a 5% water (unscented) bleach solution for at least one minute. Let dry naturally. *Do not wipe.*

COOKING

Cooking immediately post-disaster is inevitably a compromise. This is not a time to attempt gourmet meals, but those suggested (pages 85-87) are quick, healthy and nutritious.

All are easy and quick to prepare. Most need only mixing and heating tinned or dried products and use only ingredients listed. None require cooking skills.

HEAT FOR EMERGENCY COOKING

There is unlikely to be gas or electricity. There are various ways of coping – but first a few warnings.

Gas bottle barbecues should not be used. They are a major danger in cyclones and even more so in bush fires. Never use one indoors: they produce toxic carbon monoxide that can and does kill: see pages 44-45.

Burning charcoal *too* generates huge amounts of carbon monoxide. When smouldering it produces even more. It can be lethal if used indoors – even if all doors and windows are fully open.

Portable LP gas-powered cookers are safe in well ventilated areas. Twelve or so replaceable cylinders provide for two weeks cooking for a few people. The cylinders have a long shelf life but must be kept dry.

Reliable matches are essential. Use the long thick ones made for lighting barbecues. Ensure they are kept dry (in say a zip plastic bag). Don't rely on gas lighters.

COOKING

DIY SURVIVAL COOKERS

A small cooker can be made out of an old tin. Use a nail and a hammer (or stone) and bash a few holes through the bottom to allow air to enter. Use only use hardwood: softwood makes food taste bad. (Treated pine is toxic.)

A few candles give enough heat but cooking takes a fair time. A better way to use them is to melt them and use the still molten wax to saturate small, tight rolls of corrugated cardboard – and burn *those* for cooking.

If you have a big thermos add whatever you need to cook and fill it with boiling water. (Cooking takes about twice as long as normally.)

Some foods can be tightly wrapped in heavy foil and tossed onto the embers of a fire – *never* the flames.

Cook only the amount that will be eaten for one meal. Where possible, cook meals that require only one pan.

A cast iron oven can roast meat, bake bread and cook stews. They can also be placed in a hole or container filled with hot embers, with more embers on top.

FOOD COOKING SAFETY

Do not eat any under-cooked bird meat.

Particularly do not allow juices from raw birds to touch other food or preparation surfaces: especially the soft thick paper often used under frozen chicken.

All meat must be *thoroughly* cooked. Don't even think of medium-rare steak!

Throw out meat, poultry, fish, eggs or leftovers that have been above 100° C (40° F) for more than two hours. Also throw out food that has even remotely contacted flood or storm water, or has an unusual smell, color or texture.

Never taste any food to see if it 'seems' okay.

ULTRA SIMPLE MEALS

Stuff baked potatoes with a can of hot three-bean mix.

Cook a pot of rice, add refried black beans and shredded hard cheese if available.

Cook rice or pasta and pour in a can of any heated soup. Add grated hard cheese if available.

Sprinkle canned beans with grated hard cheese. Cover with bread mix. Bake for 30 minutes at medium heat.

Cook spaghetti, drain and add a couple of fried eggs. This sounds really odd but is delicious! It's better still with grated hard cheese, and also works with egg noodles.

Pour a heated can of chickpeas plus heated canned or fresh tomato, garlic (and parsley if available) over any cooked pasta.

For a more substantial dish, gently fry two cloves of garlic and a chopped onion, add a can of tomatoes, simmer for 10 minutes then add about a quart (litre) of chicken stock. If you have no stock use slightly less water. Add a can of lentils, salt and pepper (plus a pinch of cumin if available).

Once cooked, add a little spinach or anything green, simmer a few minutes more. Ideally top with grated hard cheese.

Make fritters by cubing a potato and frying until brown. Mix in a chopped onion, garlic, and any chopped veggies you can find, add to the potato and cook a further five minutes. Beat an egg or two and pour over the above. Cook for another two minutes each side.

If you end up with a total mess, scramble the lot and add shredded hard cheese. Pretend that is how it was meant to be and eat after a minute or two. It's delicious either way!

Combine canned beans, onion, garlic, spinach or anything more or less green and edible, odd bits of sliced meat or sausage plus a little cumin or oregano.

Drain and brown a can of chicken, add some dry pasta, tomato sauce, chopped onion, crushed garlic, and a can of mushrooms. Add stock or water to just cover, and then add garlic and spices. Stir, bring to the boil and simmer for a few minutes. Put the lid on,and cover with a towel or whatever for 10 to 15 minutes before serving. If really hungry eat immediately.

Nourishing soups can be made with canned vegetables (plus canned meat if required), fortified by pre-cooked, broken-up pasta.

An excellent dhal soup can be made from canned red beans, water or stock, ideally a dash of turmeric, a dash of ginger and a little garlic and onion powder.

These burgers are simpler and healthier than most. Chop an onion. Fry until cooked. Wrap it, plus a slice of cooked meat (canned or otherwise), inside two slices of buttered or 'mayonnaised' wholemeal bread or roll. Then fry (oil is not needed). Top with cheese and pickled peppers if available.

This also works well with canned tuna.

ENERGY BARS

These energy bars are nutritious and packed with the calories essential during and after disaster situations.

Break an egg into a bowl, add half a cup of brown sugar. Beat well and then add half a cup of raisins, the same of chopped nuts and about a cup of rolled oats. Stir until well mixed. Pour it into a buttered or oiled pan and flatten firmly. Bake in a pre-heated 180^0 C (360^0 F) oven or whatever you can mock up and then allow to cool.

Even better is a cup of chopped up apricots, another of chopped mixed nuts mixed with half a cup of honey and a couple of tablespoons of oil. Beat in about $^2/_3$ cup of wholemeal flour and about $^1/_2$ cup of wheat germ or oats using only enough water to form a stiff mix. A well beaten egg can be added – but is not essential.

Pour into a greased pan and bake for 25 to 35 minutes (or firm) in a 180^0 C (360^0 F pre-heated oven or whatever heating you can improvise.

COOKING

BASIC FOOD LIST

Every meal suggestion in this chapter can be made from the following:

Beans: three/five bean mix, black, lentils, chickpeas etc

Bread mix.

Butter: canned.

Canned soup.

Cheese: hard only (e.g. Parmesan or Romano).

Chicken canned.

Chicken stock powder (also beef and vegetable).

Chili sauce.

Cumin.

Eggs.

Fruit canned.

Garlic powder.

Herbs: dried including basil, oregano and parsley.

Meat: any (ground, freeze dried or canned).

Oats: rolled.

Oil (olive).

Onion (dried).

Mushrooms: canned.

Pasta.

Peanut butter.

Pickled peppers.

Potatoes.

Rice.

Sliced meat (any).

Spices including ginger and turmeric.

Tomato: dried, canned, powdered tomato.

Tomato sauce.

Tuna: canned.

Worcestershire sauce.

The so-called Ultra-high temperature processing (UHT) long life milk — packaged in sterile containers — has a typical unrefrigerated shelf life of some six to nine months). Use and replace it regularly.

PERSONAL COOKING NOTES

The simple recipes on the previous pages stem mainly from the author's own experience throughout WW2 in London, and others who have likewise been forced to cope where conventional food, power and shelter was not readily available.

It is likely that some readers have simple recipes for their own vital (or comfort) food. Do not underestimate its value. Chicken soup and energy bars etc. are huge morale boosters.

This page has been left for your own recipes. Add needed ingredients to the list on pages 88-89.

PART FIVE

General

Riots
Transport
Money
Grab bags
Checklists

RIOTS, TRANSPORT & MONEY

RIOTS

Serious riots are rare in Australia and all but unknown in New Zealand, but are not uncommon in many other countries.

The risk of deliberate personal attack is mostly low as long as you have no actual or perceived involvement. You can however be targeted if the riot is fueled by an external event.

By far the safest behavior is to stay indoors or seek refuge in a hotel and to stay well away from doors and windows.

If traveling in areas where riots are at all likely, wear clothes that cover all of your exposed skin and are not too different from the local dress. You may not fully blend in but it is best not to be overly different as rioters tend to target obvious outsiders.

There is some risk of being attacked accidentally. Never knowingly walk or drive between rival groups.

If actual rioting exists, retreat carefully out of the area. As long as you are a few hundred yards (metres) away from the action you are likely to be safe.

Carry a map of the area and know exactly where you are at all times.

If in a bus, the driver will usually pull off the road. The driver, or police, may pull down window blinds. Do not peek out: it virtually invites bullets, or rocks to be thrown at you.

Carry identification (ideally your passport) as you may be apprehended by security police. Without immediate identification there is a risk of arrest and a long wait in less than comfortable conditions.

Also, and importantly, carry the phone number of your consulate or embassy.

TRANSPORT

In many disaster situations roads are damaged and trees and rocks may partially block them. In bush fire situations road travel can be massively dangerous.

Fuel is invariably in short supply following panic buying. Even if there is fuel at a petrol station it will be unobtainable if power cuts prevent pumps working.

Storing petrol or diesel may seem a good idea but petrol tends to absorb moisture and goes stale.

Diesel fuel builds up fungi. Biocides can be used to keep fungi in check but even with ample fuel, roads are often physically impassable or closed anyway.

An old but sound conventional upright type bicycle with panniers that can carry some supplies is useful. It can be carried around obstacles and needs no fuel. Carry a stout lock and long chain so you can secure the bicycle to something immovable.

MONEY

Cash is by far the preferred currency post-disaster. It may be the *only* form usable. Power outages immobilise credit card transactions and shut down automatic teller machines.

Banks too are likely to be closed due to lack of power and communications.

Have at least the local equivalent of A$250, with most in small denominations.

GRAB BAGS

It is becoming common in the USA and Canada to prepare with extreme thoroughness against virtually apocalyptic events. Part of that preparation includes a so-called 'bug-out' or 'grab bag'.

The concept is always to carry a kit that provides immediate aid if you need to move there and then.

It's a first but major level of preparedness encompassing food, water, basic medical gear and personal identification to cope (unaided) for a typical three days.

Ready made grab bag kits are available (from the USA and Canada) but many people assemble their own.

The size and weight must be limited to what you can comfortably carry in a smallish waterproof rucksack.

Most kits contain some or all of the following.

AM/FM small battery radio plus spare battery.

Baby wipes.

Cash and change.

Clothing suitable for area and climate.

Drinking water: 5-6 litres (1-1.5 US gallons).

Duct tape and length of light terylene cord.

Fire starting: waterproof matches or ferrocerium rod.

First aid kit (basic) and first aid manual.

Folding knife: ideally a small Swiss Army knife or a Leatherman tool.

Identification: driver's license, credit card etc.

Light plastic sheet for shelter and water collection.

Light sleeping bag.

Maps of the area.

Medical records and emergency medical information book. See page 114 for recommendations.

Medication if needed, plus prescriptions.

Mobile or satellite phone.

Non-perishable food.

Paper tissues.

Pencils and paper pad.

Pre-prepared plan: emergency center location, evacuation routes, agreed meeting places.

Spectacles (spare pair).

Telephone numbers: emergency, poison control centre, friends, family, neighbors.

Toilet paper.

Torch (LED) plus spare battery.

Ultra light tent.

Walking shoes and rain gear.

Water purifying tablets.

IN CAR - (IF USABLE)
Fire extinguisher, flares, jumper leads, shovel, water and food.

CHECKLIST: THE BASICS

As stressed in this book, the aftermath of most disasters present generally similar problems. All need similar preparations – with keeping warm a priority. Exposure to cold kills more people than does lack of food and water.

Apart from keeping warm, the essentials for all are much the same. A few items are relevant only in some situations – such as possibly having an inflatable boat in flood prone areas.

To keep minds rested but occupied include a few paperbacks, playing cards, crosswords and Sudoku puzzles etc. Also include a few items that improve comfort but are not essential for health and safety. These may include small copies of family photographs and a few children's books or small toys.

SHELTER AND WARMTH

Blankets (ex army disposal stores).

Fire starters and waterproof matches.

Large, extra heavy garbage bags. Their range of uses include as emergency sleeping bags (page 60), emergency shelter, sealing broken windows etc.

Lightweight rain gear with hoods.

Mylar space blanket (also good for signalling).

Sleeping bags.

CHECKLIST: THE BASICS

Small sealable garbage bags for rubbish.

Tarpaulin with tie down provision (ideally nylon) to use as an emergency shelter or covering a damaged roof.

Warm and waterproof clothing plus a change of clothing.

FOOD AND WATER

Beverages such as coffee and tea.

Canned and other non-perishable foods not needing refrigeration, plus high calorie food bars. Also enough food for several days food that requires little preparation: see pages 83-88.

Drinking water in large sealed plastic containers – at least 2 litres (about half a gallon) per person per day. Store in a cool, dark place.

Metal water bottles.

Plain bleach for water purification.

Plastic bottles for SODIS: See pages 46-50.

Water purification tablets.

COOKING

Aluminium foil, anti-bacterial liquid.

Bucket.

Can opener (two). Reliable high quality manual can openers are essential. It's difficult and risky to open a can without one, and hard with cheap ones.

Handy wipes.

Metal grill.

Oven mitts or gloves.

CHECKLIST: THE BASICS

Paper towel, tissues.

Portable (gas) camping stove plus spare canisters.

Pots and pans.

Small and zip lock plastic bags.

HYGIENE AND PERSONAL

Detachable toilet seat.

Insect repellent.

Medications.

Soap.

Spare spectacles.

Tampons or sanitary pads.

Toilet paper (ample).

COMMUNICATIONS AND SIGNALLING

AM/FM battery radio.

Mobile phone plus recharger.

Signalling mirror.

Spotlight: high power LED (plus extra lithium batteries).

Television: small battery powered.

Whistle.

SAFETY GEAR

Dust masks (P1 or P2) See page 35 regarding P1/P2.

Fire extinguishers (vital as water may not be available).

Overalls.

Shoes with heavy soles.

Work gloves.

CHECKLIST: THE BASICS

ESSENTIAL DOCUMENTS

Contact and meeting place information for your household.

Credit cards.

Driver's licence.

Family photos and descriptions (to aid emergency personnel in finding missing people).

Financial documents.

First aid manual.

Insurance papers.

Large scale map of location area (highlight closest high ground plus agreed meeting places).

Local maps.

Map of relevant area if you have a plan to evacuate.

Medical prescriptions (copies).

Microchip numbers of pets.

Passports.

Personal identification (copies).

This book.

Wills.

TOOLS

Axe with a sheath for cold areas or machete for tropical areas.

Chisels.

Crow bar.

Fixed-blade knife or Leatherman/Swiss Army knife.

CHECKLIST: THE BASICS

Hammer.

Multipurpose tools or materials.

Pliers: normal and vice grip to turn off water/gas valves.

Screwdrivers.

Shovel.

Spade.

Wood saw.

MATERIALS (GENERAL)

Duct tape.

Glue: water resistant.

Nails: assorted.

Needle and thread.

Nuts, bolts, and screws: assorted.

Plastic bags.

Spare batteries.

Terylene cord (about 50 metres or yards).

GENERAL

Candles for light, signalling and fire starting.

Cash: a minimum of A$250 in cash with most in small denominations.

Children's toys and books (if applicable).

Clothes line.

Comprehensive first aid kit.

Extension cord/s.

CHECKLIST: THE BASICS

Garden hose (about 30 metres or 35 yards). This can also be used for emergency siphoning.

Matches (the long heavier variety are more reliable).

Spare keys for house and motor vehicles.

Special items for infant, elderly or disabled family members.

Torches, plus spare long-life batteries.

PETS
Basket or cage.

Drinking water.

Food.

Identification chip number.

Leash.

Medication.

Water dish.

TRANSPORT
Bicycle air pump.

Bicycle with luggage rack (old one with heavy duty tires).

Puncture outfit or two spare inner tubes.

OPTIONAL
Electric multi-meter (if you know how to use it).

Inflatable boat.

Test light.

Fuel for vehicles.

PART SIX

REFERENCES & DATA

BUSH FIRE WARNINGS

(Also known in some countries as brush fire or wildfire.)

Australia: myfirewatch.landgate

Canada: cwfis.cfs.nrcan.gc.ca

Europe: forest.jrc.ec.europa.eu/effis

Indonesia: indofire.landgate.wa.gov.au/indofire.asp

New Zealand: https://learn.thundermaps.com/blog-posts/fire-alerts-nz-property-owners/

USA (Continental): fsapps.nwcg.gov/afm/current.php

BUSH FIRE REFERENCES - GENERAL
abcb.gov.au/Resources/Publications/Education-Training/Private-Bushfire

Bushfire overview: http://csiro.au/en/Research/Environment/Extreme-Events/Bushfire

Complete Bushfire Safety Book (by Joan Webster)

firewisesa.org.za

https://www.ready.gov/wildfires

EARTHQUAKE WARNINGS

Seismic waves from an earthquake travel at about 190 km/minute (120 miles/minute). Because of this the most probable warning time is likely to be only seconds.

Some systems *are* operational but mainly of value (as earthquake warnings) in large organisations that can set up instant staff warning. The last two references (below) relate to actual and projected such systems.

The warnings systems now being put in place will be of huge value for Tsunami warnings - page 106.

BACKGROUND INFORMATION

earthquake.usgs.gov/research/earlywarning/background.php

earthquake.usgs.gov/research/earlywarning/nextsteps.php

seismicwarning.com/sws-home.html

techcrunch.com/2016/08/26/shakealert-provides-earthquake-early-warning-system/

www.ga.gov.au/earthquakes/

TSUNAMI WARNINGS

As tsunamis are initiated by earthquakes some of the warning sources on page 105 also provide background information.

The warning systems below are directly relevant.

Australia and off-shore territories: bom.gov.au/tsunami/about/tsunami_warnings.shtml

Japan: www.jma.go.jp/en/tsunami/

Pacific: ptwc.weather.gov/

USA: (West coast and Alaska): ntwc.arh.noaa.gov/

TSUNAMI REFERENCES

As tsunamis are initiated by earthquakes some of the references on page 105 also provide background information. The references below are directly relevant.

Australia and off-shore territories: bom.gov.au/tsunami/about/tsunami_warnings.shtml

Japan: www.jma.go.jp/en/tsunami/

Pacific: ptwc.weather.gov/

USA: (West coast and Alaska): ntwc.arh.noaa.gov/

How Tsunamis form: australiangeographic.com.au/topics/science-environment/2011/03/tsunamis-how-they-form/

Tsunamis - frequently asked questions: bom.gov.au/tsunami/info/faq.shtml

Tsunami Safety Rules: http://wcatwc.arh.noaa.gov/?page=safety

WEATHER (SEVERE) - GLOBAL

SEVERE WEATHER - GLOBAL

Africa: https://www.accuweather.com/en/africa-weather

Asia: https://www accuweather.com/en/asia-weather

Australia: bom.gov.au

Canada: weather.gc.ca/canada_e.html

China: en.weather.com.cn/

Europe: http://www.accuweather.com/en/gb/national/satellite

India: www.indiaweather.gov.in

New Zealand: metservice.com/national/home

Pacific islands: bom.gov.au/pacific/

Russia: www.accuweather.com › World › Asia

South America: weatherzone.com.au/world/south-america

USA - general: weather.gov/

Worldwide: worldweather.wmo.int/

WIND EVENT WARNINGS

Arabian Sea, Bay of Bengal: .imd.gov.in/pages/services_cyclone.php

Australia (general): http://www.bom.gov.au/cyclone/

Australia: Northern Territory: http://www.bom.gov.au/nt/

Australia: Queensland: Northern and Gulf of Carpenteria: http://www.bom.gov.au/nt/forecasts/tcoutlook.shtml

Australia: Queensland: Coral sea areas: http://www.bom.gov.au/qld/forecasts/cyclone.shtml

Australia: Western Australia: http://www.bom.gov.au/wa/

Caribbean Sea, Gulf of Mexico, North Atlantic and Eastern Pacific oceans: .nhc.noaa.gov/?epac

Central North Pacific: http://www.weather.gov/prh/

Central Pacific Ocean: http://www.prh.noaa.gov/cphc/

Coral Sea: bom.gov.au/weather/qld/cyclone.shmtl

Eastern Pacific: https://metoc.ndbc.noaa.gov/JTWC/

WIND EVENT WARNINGS

India: www.imd.gov.in/pages/allindiawxwarning
bulletin.php

Southern Hemisphere: https://metoc.ndbc.noaa.gov/
JTWC/

South-East Indian Ocean: http://severe.worldweather.
org/tc/au/

South Pacific: metservice.co.nz/public/weatherWarn-
ings/marine-warnings.shtml

South-West Pacific Ocean: www.bom.gov.au

Tasman Sea: metservice.co.nz/forecasts/high_seas.asp

WIND EVENT (TERMS)

Ten minute sustained wind km/h − m/hr	North Indian ocean	South-west Indian ocean	Australia
<52 - <32	Depression	Tropical disturbance	Tropica low
52/56 - 32/35	Deep depression	Tropical depression	As above
56/63 - 35/39	As above	As above	As above
63/89 - 39/55	Cyclonic storm	Moderate tropical storm	Tropical cyclone 1
89/119 - 55/74	Severe cyclonic storm	Severe tropical storm	Tropical cyclone 2
119/159 - 74/99	Very severe cyclonic storm	Tropical cyclone	Severe tropical cyclone 3
159/167 - 99/104	As above	As above	Severe tropical cyclone 4
167/198 - 104/123	As above	Intense tropical cyclone	Severe tropical cyclone 4
198/222 - 123/138	As above	Very intense tropical cyclone	As above
>222 - >138	Super cyclonic storm	As above	As above

WIND EVENT (TERMS)

Ten minute sustained wind km/h − m/hr	South-west Pacific	North-west Pacific South-west Indian ocean)
<52 - <32	Tropical depression	Tropical depression
52/56 - 32/35	As above	As above
56/63 - 35/39	Tropical low	As above
63/89 - 39/55	As left	Tropical storm
89/119 - 55/74	As left	Severe tropical storm
119/159 - 74/99	As left	Typhoon
159/167 - 99/104	As left	Typhoon
167/198 - 104/123	As left	Typhoon
198/222 - 123/138	Severe tropical cyclone 5	Typhoon
>222 - >138	As above	Typhoon

WIND EVENT REFERENCES

australiasevereweather.com/cyclones/global_
terminology.htm

http://www.bom.gov.au/cyclone/climatology/trends.shtml

http://www.bom.gov.au/cyclone/about/checklist.shtml

*Blueprint for Safety — comprehensive details on how
to strengthen your home:* flash.org

Preparing Your Property: https://www.dfes.wa.gov.au/
safetyinformation/cyclone/Factsheets/DFES-Cyclone_
and_Flood-Preparing_Your_Home_and_Property.pdf

*Saffir-Simpson Hurricane Wind Scale Predicts
Hurricane Wind Damage:* nhc.noaa.gov/aboutsshws.php

*Seven Things You Need to Know Before Rebuilding
Your Hurricane-Damaged Home:* flash.org/resources/
files/BFS%20News_After%20the%20Storm_Fla.pdf

REFERENCES - MEDICAL

MEDICAL (GENERAL)

Emergency First Aid: a quick guide: published by St John Ambulance Australia: stjohn.com.au

First Aid Manual. St John Ambulance. ISBN 0-7513-37048

en.hesperian.org/hhg/New_Where_There_Is_No_Doctor:Chapter_3:_First_Aid

health.qld.gov.au/mozziediseases/default.asp

Saving Lives with Emergency Medicine. Windale, Rose.

SODIS: www.sodis.ch/index_EN

REFERENCES: MEDICAL (PERSONAL)

It is of huge value in a medical emergency to have a full record of your medical history, blood type etc.

It should include full personal contact details, and also telephone numbers of ambulance, fire and police. Include also details of where to meet in an emergency.

Devise your own list - or copy the form on page 115.

VITAL MEDICAL INFORMATION
Name:
Date of birth:
Language/s spoken:
Blood type:
Medicines, dosages, how long taken:
Allergies/allergic reactions:
Date of last physical examination:
Dates/results of tests and screenings:
Major illnesses/surgeries: (and when):
Chronic (long term) diseases:
Street address:
Land line telephone number:
Mobile number:
Primary contact number:
Relative, office or school number:
Emergency contact(s):
Police:
Fire service:
Ambulance:
Where to meet in an emergency:
Complete these forms now - one for each person concerned!

REFERENCES: COMMUNICATIONS

CITIZENS BAND (CB PLUS UHF CB)
Aust: http://www.acma.gov.au/Citizen/TV-Radio/ Radio/Marine-and-Amateur-Radio/citizen-band-radio-service-cbrs-fact-sheet

https://en.wikipedia.org/wiki/UHF_CB

https://en.wikipedia.org/wiki/Citizens_band_radio

USA: https://en.wikipedia.org/wiki/CB_usage_in_the_ United_States

DIGITAL TV DEPLOYMENT WORLDWIDE
en.wikipedia.org/wiki/List_of_digital_television_ deployments_by_country

https://en.wikipedia.org/wiki/List_of_digital_ television_deployments_by_country CB

REFERENCES: GAS

https://caravanandmotorhomebooks.com/gas-risk-in-caravans/

https://www.cdc.gov/disasters/cofacts.html

https://www.cdc.gov/mmwr/preview/mmwrhtml/mm5439a7.htm

REFERENCES: GENERATOR RISK

https://www.cdc.gov/disasters/cofacts.html

https://www.cdc.gov/mmwr/preview/mmwrhtml/mm5439a7.htm

REFERENCES: GUN LAWS

Australia: http://www.loc.gov/law/help/firearms-control/australia.php

Canada: https://en.wikipedia.org/wiki/Gun_laws_in_Canada

Global: en.wikipedia.org/wiki/Overview_of_gun_laws_by_nation

USA: https://en.wikipedia.org/wiki/Gun_laws_in_the_United_States_by_state

REFERENCES - WATER

SODIS: www.sodis.ch/index_EN

A Guide to Commercially-Bottled Water and Other Beverages: https://www.cdc.gov/parasites/crypto/gen_info/bottled.html

A Guide to Drinking Water Treatment and Sanitation for Backcountry and Travel Use: www.cdc.gov/parasites/crypto/gen_info/filters.html/ (This guide is also available in PDF)

A Guide to Water Filters: www.cdc.gov/parasites/crypto/gen_info/filters.html

ABOUT THE AUTHOR

ABOUT THE AUTHOR

As a ten year old child and then young teenager, Collyn's circumstances were such that he had no choice but to cope alone in London throughout World War Two.

He joined the Royal Air Force when 17, and trained as a ground radar engineer. He later became a research engineer. http://vauxpedianet.uk2sitebuilder.com/vauxhall---chaul-end-engineering-research-test-centre

From 1959-1960, Collyn drove a 4WD truck (twice) the length and breadth of Africa. He then migrated to Australia, where, he founded *Electronics Today International,* (acclaimed Best Electronics Publication in the World (by Union International de la Presse Radiotechnique et Electronique in 1976. https://en.wikipedia.org/wiki/Electronics_Today_International

In 2000, he and his Finnish-born now psychologist wife (Maarit) moved to remote north-west Australia.

There, they built a cyclone-proof home in a remote area lacking all but bore water. They built a big solar system to provide power for building and later the house and food-generating irrigation.

Whilst there, they coped with the worst cyclone to hit the area in 100 years. In 2010, they moved to a (now all-solar) house in Church Point, Sydney.

ACKNOWLEDGEMENTS

The concept of this book originated from a business acquaintance who suggested my 'unusual' background plus writing/publishing experience would provide the insight required.

I thank my daughter, Fiona Rivers BA (Hons) for turning it into more elegant prose, and my grandson Max Nolan for the original design and cover concept.

I also thank my (child-psychologist wife) Maarit Rivers BA, MA, for textual suggestions, proof reading and looking after home and business during the intensity of writing and production.